D0761891

Magnetic Resonance Imaging

PHYSICAL PRINCIPLES AND
APPLICATIONS

This is a volume in

ELECTROMAGNETISM

ISAAK MAYERGOYZ, SERIES EDITOR
UNIVERSITY OF MARYLAND
COLLEGE PARK, MARYLAND

Magnetic Resonance Imaging

PHYSICAL PRINCIPLES AND
APPLICATIONS

VADIM KUPERMAN

University of Chicago
Chicago, Illinois

ACADEMIC PRESS

A Harcourt Science and Technology Company

San Diego   San Francisco   New York   Boston
London   Sydney   Tokyo

ACADEMIC PRESS
A Harcourt Science and Technology Company
525 B Street, Suite 1900, San Diego, CA 92101-4495, USA
http://www.apnet.com

Academic Press
Harcourt Place, 32 Jamestown Road, London NW1 7BY UK
http://www.hbuk.co.uk/ap/

Library of Congress Catalog Card Number: 99-65743
International Standard Book Number: 0-12-429150-3

Printed in the United States of America
00 01 02 03 IP 9 8 7 6 5 4 3 2 1

To my dear wife Leslie,
whose constant love and support are precious to me.
I could not have written this book without her by my side.

Contents

Foreword

This volume in the Academic Press Electromagnetism series presents a self-contained introduction to magnetic resonance imaging (MRI). This technology has been evolving at a remarkable pace and it has become a very valuable and even indispensable diagnostic tool in various areas of medicine. This remarkable expansion of MRI technology has been achieved as a result of coordinated progress in such areas as physics of nuclear magnetic resonance, invention of novel image acquisition and image reconstruction techniques and development of sophisticated MRI instrumentation. Thus, it is becoming increasingly important to give the exposition of MRI technology with the emphasis on its interdisciplinary nature. This is exactly what the author of this book has accomplished.

This is a short and concise book that nevertheless covers an extraordinary amount of technical information. It is intended for readers without extensive experience in the area of MRI technology. It reviews the basic underlying physical principles of nuclear magnetic resonance by and large within the framework of classical electromagnetism. The book also covers the main MRI imaging techniques with the emphasis on such topics as image contrast, signal-to-noise ratio, image artifacts, rapid imaging, and flow imaging. The final chapter of the book contains a brief discussion of MRI instrumentation.

I believe that this book will be very attractive as a clear and concise introduction to the MRI technology. As such, it will be a valuable reference for beginners and practitioners in the field. Physicists, radiologists, electrical and biomedical engineers will find this book very informative.

Isaak Mayergoyz, Series Editor

Acknowledgements

I would like to thank Drs. Isaak Mayergoyz and Zvi Ruder who were instrumental in creating the initial impetus for the writing of this book and were very supportive during the writing process. Throughout my work on the manuscript, I benefited from discussions with my many colleagues at the University of Chicago. In particular I owe a great deal to Drs. David Levin and David Chu. My special gratitude goes to Drs. Marcus Alley, David Chu and Gary Friedman who read parts of the manuscript and made very insightful comments. Lastly, I am greatly indebted to Dr. Leslie Lubich for her invaluable help in editing the manuscript.

Vadim Kuperman, Ph.D.

Introduction

Since its first implementation by Lauterbur [1], Magnetic Resonance Imaging (MRI) has become an important noninvasive imaging modality. MRI has found a number of applications in the fields of biology, engineering, and material science. Because it provides unique contrast between soft tissues (which is generally superior to that of CT) and high spatial resolution, MRI has revolutionized diagnostic imaging in medical science. An important advantage of diagnostic MRI as compared to CT is that the former does not use ionizing radiation.

MRI is based on the phenomenon of Nuclear Magnetic Resonance (NMR), independently discovered by Bloch *et al.* [2] and Purcell *et al.* [3]. Although the potential of NMR spectroscopy was recognized almost immediately and its development started soon after the discovery of NMR, it took more than two decades to implement NMR-based imaging. Even after the pioneering work by Lauterbur and the development of basic imaging techniques by Kumar *et al.* [4] and Mansfield [5], several more years were required to design and develop imaging hardware at the level necessary to produce high-quality diagnostic images of the human body. Despite its relatively slow beginning, MRI has become an indispensable diagnostic tool since the early 1980s.

The foundation of the NMR phenomenon is the interaction between an external magnetic field and nuclei which have a nonzero magnetic moment. According to the classical theory of electromagnetism, the motion of individual nuclear moments in a static magnetic field B_0 is a precession about B_0 at an angular frequency ω_0, known as the *Larmor frequency*, which is proportional to the strength of the

1

magnetic field. Another well-known fact is that the energy of inter-action with B_0 depends on the direction of nuclear moments in such a way that the minimum energy corresponds to the state in which the moments are parallel to B_0. As a result, in thermal equilibrium the majority of nuclear magnetic moments are aligned along the external field. The alignment of magnetic moments gives rise to non-zero magnetization in macroscopic samples of solids, liquids, or gases containing a large number of nuclei (e.g., $\sim 10^{23}\,\mathrm{cm}^{-3}$ of hydrogen nuclei in water).

The NMR phenomenon is observed when a macroscopic sample in a static magnetic field is irradiated by an oscillating magnetic field of frequency ω that equals the frequency of precession, ω_0. The NMR phenomenon can be best explained by using the arguments put forward by Bloch [6]. Suppose that a macroscopic sample is placed between the poles of a magnet that produces a static magnetic field, B_0. Under the influence of B_0 the sample becomes magnetized. In thermal equilibrium the nuclear magnetization in the sample can be expressed as

$$M = \chi B_0, \tag{1}$$

where χ is the nuclear susceptibility. Suppose that the sample experiences an oscillating magnetic field B_1 produced by an alternating current in a coil encompassing the sample, and assume that the oscillating field is perpendicular to B_0. It can be shown that at resonance, achieved when the frequency of B_1 equals ω_0, even a weak oscillating field applied as a pulse can rotate the magnetization in the sample and place it in the transverse (i.e., perpendicular to B_0) plane.[1] After the excitation pulse ends, the transverse magnetization in the sample precesses about B_0. During the precession the transverse magnetization decays because of nuclear interactions and nonuniformity of B_0 (see Chapter 1). From Faraday's law of induction it follows that the time-varying magnetization induces voltage in the coil. The induced voltage, which identifies the presence of the transverse magnetization in the sample, can be Fourier transformed to obtain a NMR spectrum. For many applications of NMR spectroscopy it is critically important that the observed signal includes contributions from nuclei in different chemical environments (e.g., hydrogen nuclei in water and fat). Such nuclei normally have slightly different frequencies of precession depending on the chemical composition of a sample. NMR spectra

---

[1]In contrast, off resonance excitation normally produces negligibly small components of the magnetization in the transverse plane (see Chapter 1).

can be used to identify chemically different populations of nuclei and to determine their relative amounts in a sample.

We are indebted to F. Bloch for a set of widely used phenomenological equations that describe the dynamics of nuclear magnetization in media [6]. In particular, Bloch has predicted that thermal fluctuations would cause exponential relaxation of the longitudinal (i.e., parallel to B_0) magnetization, M_{lon}, to the equilibrium state in equation (1):

$$M_{lon} - \chi B_0 \propto \exp(-t/T_1), \tag{2}$$

where T_1 is a characteristic relaxation time.

On the other hand, Bloch has suggested that magnetic interactions between neighboring nuclei cause exponential decay of the transverse (i.e., perpendicular to B_0) magnetization, M_{tr}, with a characteristic time constant T_2:

$$M_{tr} \propto \exp(-t/T_2). \tag{3}$$

Although the Bloch model has important limitations (e.g., it fails to describe NMR in solids), it has proven to work well for liquids and biological systems in general. Throughout this book the Bloch equations (discussed in more detail in Chapter 1) are used as the main tool for analysis of the NMR phenomenon and associated effects important for MRI.

The pioneering work of Lauterbur [1] has demonstrated that the NMR signal acquired in the presence of external magnetic field gradients can be used to obtain MR images of the transverse magnetization in the object. The basic effect that makes MRI possible is that in the presence of magnetic field gradients the frequency of precession, ω_0, becomes spatially dependent. The observed NMR signal is a sum of many signals produced by nuclei at different locations in the sample. Each component of the signal acquired in the presence of magnetic field gradients is characterized by its unique frequency and phase. How can we produce an image from the signal that is given by a sum of different frequency components? To answer this question we need to review the basic concepts of spatial encoding and image reconstruction in MRI.

It can be shown (see Chapter 3) that the fundamental relationship between the NMR signal, S, and the transverse magnetization in the imaged object can be expressed as

$$S(k) \propto \int M_{tr} e^{jkr} \, dV. \tag{4}$$

In this equation k is a spatial frequency defined by external magnetic field gradients used for imaging, and the integral is taken over the object's volume. This equation is extremely important for MRI because it shows that the signal is given by Fourier transform of transverse magnetization in the object. At this stage we need to recall the well-known property of Fourier transform: if $\hat{f}(k)$ is the Fourier transform of $f(x)$, then $f(x)$ is the inverse Fourier transform of $\hat{f}(k)$. After examining equation (4), it becomes apparent that an image of the transverse magnetization in the object can be produced by computing inverse Fourier transform of the signal, S.

In general, spatially encoded signals in MRI are obtained by using a number of repetitive excitations of the nuclear magnetization in the object. After each excitation the NMR signal is sampled a number of times (typically 256) during a short acquisition interval (limited by the decay of the transverse magnetization) in the presence of external magnetic field gradients. Most of the complexity in the reconstruction of MR images exists due to the fact that in practice the continuous NMR signal can only be represented by a series of its discreet samples. The problem of reconstructing a continuous function from its discreet samples is well known in signal processing. In particular, the famous *sampling theorem* [7] defines conditions under which a continuous function can be completely recovered from its samples. Principles of signal sampling in MRI as well as the relationship between imaging parameters (including strength and duration of imaging gradients, number and interval between excitations, etc.) and such important factors as scan time, image resolution, image contrast, and signal-to-noise ratio are discussed in Chapters 3 through 5.

Hydrogen nuclei (^{1}H) appear to be the best target for *in vivo* MRI. They have the following advantages: a) among all nuclei present in tissues hydrogen nuclei produce the greatest NMR signal; b) ^{1}H MRI *in vivo* achieves excellent contrast between different tissues. The remarkable feature of ^{1}H MRI is that it provides a variety of ways to manipulate image contrast. Since the inception of MRI it has been apparent that the major source of contrast in MR images is the dependence of T_1 and T_2 relaxation times, and proton density on the biochemical composition of different tissues. The studies of Damadian [8] and other investigators [9] have demonstrated that in certain instances malignant tissues have longer T_1 and T_2 relaxation times than normal tissues. Because of the exponential dependence on the tissue relaxation times, the NMR signal is extremely sensitive to small variations in T_1 and T_2. Moreover, by adjusting acquisition parameters (see Chapter 4) the dependence of the signal on the relaxation times can easily be varied in order to improve image contrast. For many applications of MRI it is important that contrast between different tissues can be further enhanced by using certain materials that can significantly shorten T_1 and T_2 in tissue. It is noteworthy that in some cases administered materials are preferentially absorbed by malignant tissues making it possible to improve identification of malignancies in MR images. Different mechanisms of contrast as well as options for manipulating contrast in MRI are discussed in Chapter 4.

When acquiring MR images the major concern is the relative amount of noise present in the signal. The signal-to-noise ratio (SNR) is one of the most important image characteristics because it defines the observer's ability to distinguish between different structures in an image. Noise in NMR can be considered as a component of the signal produced by randomly fluctuating currents in a receiver coil and imaged object. It is interesting that the relative importance of these two sources of noise depends upon the strength of the applied magnetic field: at low field strengths the noise from the coil dominates; at high field strengths the noise from the object becomes more important (see Chapter 5). In principle, the signal-to-noise ratio in MRI depends on two different kinds of imaging parameters: *intrinsic* parameters (e.g., T_1, T_2, proton density), which are typically beyond our control; and user-controlled parameters (e.g., spatial resolution, acquisition time, and signal averaging), which can vary depending on the goals of a particular MRI study. For example, in many instances high spatial resolution might be necessary to identify small structures in images. On the other hand, since the SNR in MR images decreases as spatial resolution increases, the quality of images acquired with high spatial resolution might be very poor. As a result MR imaging frequently entails a compromise between the conflicting requirements of high spatial resolution and high SNR. Image SNR can normally be increased by using signal averaging. The main limitation of signal averaging is that it increases scan time in MRI. Moreover, signal averaging is relatively ineffective because the SNR increases slowly with the number of times the signal is averaged. Therefore, it is important to know if and how SNR can be increased by varying other imaging parameters (e.g., matrix size and acquisition time). The SNR dependence on basic imaging parameters and sampling schemes used for acquisition of MR data is discussed in Chapter 5.

A well-known problem in MRI is the existence of various artifacts that can significantly degrade MR images and can mistakenly be identified as real structures in the object. Artifacts in MRI are most often caused by magnetic field inhomogeneities and different physiologic motions of patients during scanning.

Despite the fact that when designing and manufacturing MR scanners special care is taken to ensure a high degree of homogeneity of the main magnetic field B_0, in practice magnetic field inhomogeneities are always present due to spatial variations in the magnetic susceptibility of the object. Magnetic field inhomogeneities cause two different kinds of artifacts in MR images: geometric distortion and signal loss. Geometric distortion results from the presence of intrinsic magnetic field gradients that alter the dependence of Larmor frequencies of nuclear spins on their spatial locations. Signal loss is caused by dephasing of nuclear spins in a nonuniform magnetic field. Image distortions and signal loss can vary from mild to severe depending on the degree of magnetic field nonuniformity, imaging parameters, and technique used to acquire MR images. A detailed discussion of these MRI artifacts is presented in Chapter 6.

One of the fundamental assumptions of MRI is that the intrinsic dynamics of transverse magnetization in the object does not change during image acquisition. In reality this assumption can often be violated as a result of different motions in the imaged system of nuclei. Examples of such motions, important for *in vivo* MRI, include blood flow, cardiac and respiratory motions. Motion-induced artifacts can vary in appearance. One example is a series of equally spaced artificial structures produced as a result of periodic motion (e.g., blood flow pulsations). Another example is image blurring caused by the motion of the object as a whole. Chapter 6 contains a discussion of different motion artifacts and various techniques used to reduce these artifacts in MR images.

Analysis of spatial encoding in MRI as well as practical experience clearly indicate that motion artifacts can be reduced by shortening scan time. Historically, the need to decrease motion artifacts was one of the main factors responsible for the development of rapid MRI. It is interesting that one of the fastest acquisition schemes in MRI, known as *echo-planar imaging*, was proposed by Mansfield soon after the inception of MRI [5]. However, it took more than a decade to develop special hardware and to improve the data acquisition scheme in echo-planar imaging before it became widely used. The main advantage of echo-planar imaging over the conventional approach is that the former, in principle, allows acquisition of a whole image after a single excitation of the transverse magnetization in the object, making it possible to obtained echo-planar images in a fraction of a second. Such short scan times are achieved at the expense of prolonged signal acquisition. Unfortunately, increased decay of the transverse magnetization resulting from long acquisition time often causes significant artifacts in echo-planar imaging. An alternative approach for rapid MRI (based on a more conventional acquisition scheme) achieves short scan times by reducing the interval between successive r.f. excitations. The latter approach has become commonly used because it does not require special hardware and imposes less stringent requirements on magnetic field uniformity than echo-planar imaging. This and other techniques for rapid MRI are considered in Chapter 7.

Because this book is intended primarily for those without extensive experience in MRI, its main goal is for the reader to develop an understanding of MRI and its applications. To achieve this goal it seems appropriate to begin with a review of the basic physical principles of NMR, including relaxation of magnetization in media, spin echoes, and attenuation of the NMR signal due to diffusion. In Chapter 1, discussion of these subjects is based on the classical theory of electromagnetism with the addition of certain concepts from quantum mechanics useful for understanding the NMR phenomenon. Excitation of nuclear magnetization in a sample is the subject of Chapter 2. Chapter 3 considers principles of spatial encoding and the conventional acquisition scheme used in MRI. The material from the first three chapters is used in Chapters 4 through 8, which discuss particular aspects and applications of MRI including image contrast, signal-to-noise ratio, image artifacts, techniques for rapid MRI, and flow imaging. Finally, Chapter 9 contains a discussion of the main components of MRI instrumentation including magnets, gradient coils and radio-frequency coils.

References

[1] P.C. Lauterbur. "Image formation by induced local interactions: examples employing nuclear magnetic resonance," *Nature* **242**, 190 (1973).

[2] F. Bloch, W.W. Hansen, M. Packard. "Nuclear induction," *Phys. Rev.* **69**, 127 (1946).

[3] E.M. Purcell, H.C. Torrey, R.V. Pound. "Resonance absorption by nuclear magnetic moments in a solid," *Phys. Rev.* **69**, 37 (1946).

[4] A. Kumar, D. Welti, R.R. Ernst. "NMR Fourier Zeugmatography," *J. Magn. Reson.* **18**, 69 (1975).

[5] P. Mansfield. "Multi-planar image formation using NMR spin echoes," *J. Phys. C* **10**, L55 (1977).

[6] F. Bloch. "Nuclear induction," *Phys. Rev.* **70**, 460 (1946).

[7] R.N. Bracewell. *The Fourier transform and its applications.* McGraw-Hill (1986).

[8] R. Damadian. "Tumor detection by nuclear magnetic resonance," *Science* **171**, 1151 (1971).

[9] P.A. Bottomley, C.J. Hardy, R.E. Argersinger, G. Allen-Moore. "A review of ^{1}H nuclear magnetic resonance relaxation in pathology: are T_1 and T_2 diagnostic?" *Med. Physics* **14**, 1 (1987).

CHAPTER 1

Basic Principles of Nuclear Magnetic Resonance

1.1. THE PHENOMENON OF NMR

The phenomenon of Nuclear Magnetic Resonance (NMR) was independently discovered by two groups of physicists headed by F. Bloch and E.M. Purcell [1–4]. The basic physical effect at work in NMR is the interaction between nuclei with a nonzero magnetic moment and an external magnetic field. The NMR phenomenon is observed when a system of nuclei in a static magnetic field $\mathbf{B}_0$ experiences a perturbation by an oscillating magnetic field. The frequency ω of the oscillating field must satisfy the following condition:

$$\hbar\omega = |E_i - E_{i'}|. \qquad (1.1.1)$$

In this equation E_i and $E_{i'}$ are two Zeeman energies of the magnetic interaction between a nucleus and $\mathbf{B}_0$. To determine the resonant frequency in (1.1.1), recall that the interaction is described by the Hamiltonian

$$H = -\mathbf{\mu}\mathbf{B}_0, \qquad (1.1.2)$$

where $\mathbf{\mu}$ is the magnetic moment of a nucleus. The moment $\mathbf{\mu}$ can be expressed as

$$\mathbf{\mu} = \gamma\hbar\mathbf{I}, \qquad (1.1.3)$$

where $\hbar\mathbf{I}$ is the angular momentum of the nucleus and γ is a constant called the *gyromagnetic ratio*. The energy of interaction in (1.1.2)

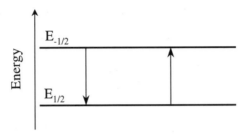

Figure 1.1. Zeeman energies for the case $I = 1/2$.

depends on the direction of the magnetic moment relative to the applied field. For example, in the case when nuclei have spin 1/2 (see Figure 1.1), the allowed energies are given by

$$E = \begin{cases} -\gamma\hbar B_0/2, & \text{if } \boldsymbol{\mu} \text{ is parallel to } \mathbf{B}_0 \\ \gamma\hbar B_0/2, & \text{if } \boldsymbol{\mu} \text{ is anti-parallel to } \mathbf{B}_0. \end{cases} \qquad (1.1.4)$$

In a general case the Zeeman energies can be expressed as $E_m = -m\gamma\hbar B_0$, where $m = -I, -I+1, \ldots, I$. From the perturbation theory it follows that an alternating magnetic field can cause transitions between states m and m', only if $m - m' = \pm 1$ [5]. In other words, only transitions between adjacent Zeeman levels separated by energy $\delta E = \gamma\hbar B_0$ are permitted. This result makes it clear that the resonant frequency in Eq. (1.1.1) is given by

$$\omega = \delta E/\hbar = \gamma B_0. \qquad (1.1.5)$$

It should be noted that for typical magnetic field strengths used in NMR imaging and spectroscopy ($\sim$1 Tesla), the resonant frequencies belong to the r.f. region of the electromagnetic spectrum (see Table 1.1).

In NMR experiments an applied r.f. field induces transitions between adjacent Zeeman energy levels. As an example, we consider hydrogen nuclei with spin $I = 1/2$. In thermal equilibrium the probability of finding a nucleus in a particular state with energy E_m is given by

$$P_m = \frac{\exp(-E_m/kT)}{Z}, \qquad (1.1.6)$$

where $Z = \sum_m \exp(-E_m/kT)$ is known as the partition function, k is the Boltzmann constant and T is the temperature. Because $E_{1/2} < E_{-1/2}$, the equilibrium population of hydrogen nuclei $(n_{1/2})$ with the lower energy $E_{1/2}$ exceeds the population of nuclei $(n_{-1/2})$ with the higher

Table 1.1. NMR constants for several nuclei used in biological applications of NMR

Element	Spin	Gyromagnetic ratio ($s^{-1} T^{-1}$)	Natural abundance (%)	Resonance frequency, $\nu = \omega/2\pi$ (MHz) at 1 Tesla
^{1}H	1/2	2.675×10^8	99.98	42.573
^{13}C	1/2	6.726×10^7	1.11	10.705
^{19}F	1/2	2.517×10^8	100.0	40.052
^{23}Na	3/2	7.077×10^7	100.0	11.263
^{31}P	1/2	1.083×10^8	100.0	17.237

energy $E_{-1/2}$. Consequently, there are more induced transitions from the lower energy level to the higher energy level than the reverse. Although the difference in population of the energy levels is very small (e.g., $(n_{1/2} - n_{-1/2})/n_{1/2} \approx \gamma \hbar B_0/kT \sim 10^{-5}$ at 1 Tesla), the large number of nuclei participating in the induced transitions between these levels makes it possible to observe absorption of r.f. energy in macroscopic samples of solids, liquids or gases. Since the probability of induced transitions reaches its maximum when the r.f. frequency reaches γB_0 [5], we can expect maximum energy absorption at the resonant frequency.

Purcell, Torrey and Pound [4] studied absorption of r.f. energy in paraffin containing hydrogen nuclei. In their classic experiment the frequency ω and amplitude of an oscillating field $\mathbf{B}_1$ were fixed while the strength of a static magnetic field $\mathbf{B}_0$, applied perpendicular to $\mathbf{B}_1$, was gradually varied. A sharp peak in the absorption spectrum was indeed observed when the strength of $\mathbf{B}_0$ reached ω/γ.

We shall see later in this chapter that the NMR signal from a macroscopic sample is proportional to the equilibrium magnetization in the sample $\mathbf{M} = n_I \langle \boldsymbol{\mu} \rangle$, where n_I and $\langle \boldsymbol{\mu} \rangle$ are the concentration and average magnetic moment of nuclei, respectively. Because the energy of magnetic interaction in (1.1.2) reaches its minimum when $\boldsymbol{\mu}$ is parallel to $\mathbf{B}_0$, the probability that a nuclear magnetic moment is parallel to $\mathbf{B}_0$ is greater than its being antiparallel to $\mathbf{B}_0$. Consequently in thermal equilibrium more magnetic moments in a sample are aligned with the external field than against the field. On the other hand, because the energy of interaction does not depend on the component of $\boldsymbol{\mu}$ in the plane perpendicular to $\mathbf{B}_0$ (referred to as the *transverse plane*), the average transverse component of $\boldsymbol{\mu}$ is zero. It is therefore clear that the average nuclear moment $\langle \boldsymbol{\mu} \rangle$ is parallel to the applied field. Because **M** is proportional to $\langle \boldsymbol{\mu} \rangle$, the equilibrium magnetization satisfies the following conditions:

$$M_z = M_0 > 0, \quad M_x = 0 \quad \text{and} \quad M_y = 0, \tag{1.1.7}$$

where the z-axis is chosen in the direction of $\mathbf{B}_0$. The equation describing magnetization in a system of nuclei with a spin I and gyromagnetic ratio γ_I can be written as

$$M_0 = \frac{n_I \hbar^2 \gamma_I^2 I(I+1)}{3kT} B_0, \tag{1.1.8}$$

assuming that the energy difference between adjacent Zeeman levels is much smaller than thermal energy of nuclei, that is, $\hbar \gamma_I B_0 / kT \ll 1$ [5]. Note that this condition is normally satisfied in NMR spectroscopy and imaging experiments.

As we will see in Section 1.4, the observed NMR signal is created by the transverse component of nuclear magnetization, $\mathbf{M}_{tr}$ ($\mathbf{M}_{tr} \perp \mathbf{B}_0$), in a sample. It is therefore important to verify whether the transverse magnetization is indeed excited by a r.f. field applied in the presence of a static field $\mathbf{B}_0$. We can prove excitation of $\mathbf{M}_{tr}$ in the case when an r.f. field $\mathbf{B}_1$ is perpendicular to $\mathbf{B}_0$ by using the equation for the absorbed r.f. power P:

$$P = - \int_{sample} \overline{\mathbf{M} \frac{d\mathbf{B}_1}{dt}} \, dV. \tag{1.1.9}$$

Since it has been established that P is nonzero during r.f. irradiation [4], we can deduce that the transverse component of magnetization must somehow be excited by the irradiating field. We shall consider the dynamics of spins during excitation in the following section.

1.2. MOTION OF MAGNETIC MOMENTS

Although only quantum mechanics can completely describe the NMR phenomenon, some of its features can be explained by the classical theory of electromagnetism. In the classical theory the motion of a magnetic moment $\boldsymbol{\mu}$ in an external field $\mathbf{B}_0$ is described by the following equation:

$$d\boldsymbol{\mu}/dt = \gamma \boldsymbol{\mu} \times \mathbf{B}_0 \tag{1.2.1}$$

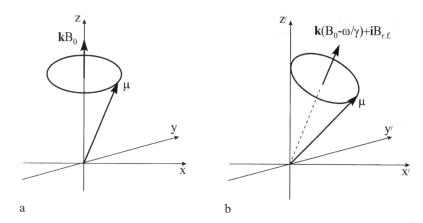

Figure 1.2. (a) Precession of the magnetic moment **μ** about **B**$_0$ in the laboratory coordinate system; (b) precession of **μ** about the effective magnetic field $\mathbf{k}(B_0 - \omega/\gamma) + \mathbf{i}B_{\text{r.f.}}$ in the rotating coordinate system.

In the case of static **B**$_0$, a solution of the last equation can be written as

$$\mu_x(t) = \mu_x(0) \cos \omega_0 t + \mu_y(0) \sin \omega_0 t, \tag{1.2.2a}$$

$$\mu_y(t) = \mu_y(0) \cos \omega_0 t - \mu_x(0) \sin \omega_0 t, \tag{1.2.2b}$$

$$\mu_z(t) = \mu_z(0), \tag{1.2.2c}$$

where $\omega_0 = \gamma B_0$ and the z-axis is chosen along **B**$_0$. According to (1.2.2a–c) **μ** precesses about **B**$_0$ on the surface of a cone at a Larmor frequency γB_0 (see Figure 1.2(a)).

In the presence of an r.f. field **B**$_1$ of frequency ω, the motion of **μ** in the laboratory coordinate system is described by the equation

$$d\boldsymbol{\mu}/dt = \boldsymbol{\mu} \times \gamma(\mathbf{B}_0 + \mathbf{B}_1). \tag{1.2.3}$$

To obtain a solution to Eq. (1.2.3) we consider a coordinate system rotating around the z-axis with an angular frequency $-\omega$. The relationship between the components of an arbitrary vector **a** in the laboratory coordinate system (a_x, a_y, a_z) and the rotating reference frame $(a_{x'}, a_{y'}, a_{z'})$ is given by

$$a_{x'} = a_x \cos \omega t - a_y \sin \omega t, \quad a_{y'} = a_x \sin \omega t + a_y \cos \omega t, \quad a_{z'} = a_z. \tag{1.2.4}$$

We now wish to consider a case when the field $\mathbf{B}_1$ is linearly polarized and perpendicular to $\mathbf{B}_0$. That is,

$$B_{1,x} = B_1 \cos \omega t, \quad B_{1,y} = 0, \quad B_{1,z} = 0. \qquad (1.2.5)$$

The linearly polarized magnetic field in Eq. (1.2.5) can be considered as the sum of two circularly polarized components rotating around the z-axis at the same frequency but in opposite directions. We shall focus on the effect of the component rotating in the direction of the Larmor precession in the laboratory coordinate system, because the counterrotating component only slightly perturbs the motion of nuclear magnetic moments and can therefore be neglected. By using Eqs. (1.2.3)–(1.2.5) we obtain the following equation of motion in the rotating reference frame:

$$d\boldsymbol{\mu}/dt = \boldsymbol{\mu} \times \gamma[\mathbf{k}(B_0 - \omega/\gamma) + \mathbf{i}B_{\mathrm{r.f.}}], \qquad (1.2.6)$$

where $\mathbf{i}$ and $\mathbf{k}$ are the unit vectors in the x' and z' directions in the rotating frame of reference, $B_{\mathrm{r.f.}} = B_1/2$. According to Eq. (1.2.6) the dynamics of $\boldsymbol{\mu}$ is defined by the effective *static* magnetic field $\mathbf{B}_{\mathrm{eff}} = \mathbf{k}(B_0 - \omega/\gamma) + \mathbf{i}B_{\mathrm{r.f.}}$. Therefore, in the rotating reference frame, $\boldsymbol{\mu}$ precesses about $\mathbf{B}_{\mathrm{eff}}$ at an angular frequency $\gamma[(B_0 - \omega/\gamma)^2 + B_{\mathrm{r.f.}}^2]^{1/2}$ (Figure 1.2(b)).

Equation (1.2.6) greatly simplifies the analysis of the motion of $\boldsymbol{\mu}$ by eliminating the time dependence of the r.f. field. Notice that at resonance $(\omega = \gamma B_0)$ the effective magnetic field in the rotating reference frame equals $\mathbf{i}B_{\mathrm{r.f.}}$. To consider the effect on resonance r.f. irradiation we assume that initially (i.e., before the beginning of the irradiation) $\boldsymbol{\mu}$ is parallel to the static field $\mathbf{B}_0$. During the irradiation the magnetic moment will precess about $\mathbf{i}B_{\mathrm{r.f.}}$ in the y'–z' plane in the rotating reference frame at an angular frequency $\gamma B_{\mathrm{r.f.}}$. For example, a 90 degree rotation of the magnetic moment will be completed at time $t = \pi/2\gamma B_{\mathrm{r.f.}}$ when the moment is along the y'-axis. If the r.f. irradiation ends immediately after the completion of the 90 degree rotation, then $\boldsymbol{\mu}$ will precess about $\mathbf{B}_0$ in the transverse plane in the laboratory coordinate system. Such excitation is known as a 90 degree pulse. An excitation pulse, which rotates $\boldsymbol{\mu}$ by 180 degrees in the rotating reference frame, is referred to as a 180 degree pulse. Notice that at resonance even a relatively weak r.f. field can cause rotation by an arbitrary angle in the y'–z' plane.

Because nuclear magnetization $\mathbf{M}$ is proportional to the average magnetic moment $\langle \boldsymbol{\mu} \rangle$, Eq. (1.2.6) can be used to describe the dynamics of $\mathbf{M}$ in a sample under the assumption that internuclear interactions during r.f. irradiation can be neglected. Based on the observation that

$\mathbf{B}_{\text{eff}}$ makes an angle θ with $\mathbf{B}_0$ such that $\tan\theta = B_{\text{r.f.}}/(B_0 - \omega/\gamma)$, we find that in the case when $B_{\text{r.f.}} \ll |B_0 - \omega/\gamma|$ nuclear magnetization initially parallel to $\mathbf{B}_0$ will remain aligned along $\mathbf{B}_0$ with very small transverse components of $\mathbf{M}$ produced during excitation. It then follows that the resonance condition $\omega = \gamma B_0$ is necessary for effective excitation of the transverse magnetization by a relatively weak r.f. field $B_{\text{r.f.}} \ll B_0$ normally used in NMR imaging and spectroscopy.

1.3. THE BLOCH EQUATIONS

The classical model of motion of a free magnetic moment (see previous section), although useful for understanding of the NMR phenomenon, cannot explain many of its important features defined by the interactions between nuclei. To overcome some of the limitations of the classical model, F. Bloch introduced in 1946 the phenomenological equations describing the dynamics of nuclear magnetization which have become an extremely useful tool for theoretical analysis in NMR imaging and spectroscopy.

The foundation of the Bloch model [3] can be understood by considering relaxation of nuclear magnetization $\mathbf{M}$ in a sample after an excitation r.f. pulse. We already know that in thermal equilibrium, $\mathbf{M}$ lies in the direction of an external static field $\mathbf{B}_0$. Bloch suggested that the equilibrium state is established because of two different processes governing the dynamics of $\mathbf{M}$: thermal perturbations and internuclear interactions. In the Bloch model [3] it is the thermal perturbations which cause relaxation of the longitudinal magnetization M_z (assuming that the z-axis is taken along the direction of $\mathbf{B}_0$) to its equilibrium state with the minimum energy of spins $E = -\mathbf{B}_0\mathbf{M}$, while interactions between nuclei in the sample cause decay of the transverse magnetization $\mathbf{M}_{\text{tr}} = \{M_x, M_y\}$ without affecting E.

According to reference [3] the relaxation of M_z is described by the equation

$$dM_z/dt = (M_0 - M_z)/T_1, \tag{1.3.1}$$

where T_1 is a constant known as the *spin–lattice relaxation time*. A solution of this equation can be written as

$$M_z = M_0 + [M_z(0) - M_0]\exp(-t/T_1), \tag{1.3.2}$$

where $M_z(0)$ is the initial value of M_z. Relaxation of M_z after a 90 degree pulse is shown in Figure 1.3.(a).

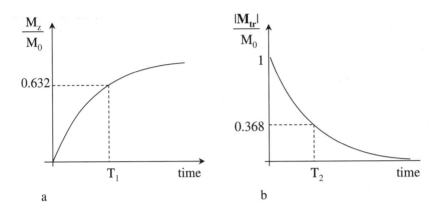

Figure 1.3. (a) Relaxation of M_z after a 90 degree pulse; (b) decay of $\mathbf{M}_{tr}$ after a 90 degree pulse.

In a system of identical noninteracting nuclei placed in a static uniform magnetic field, nuclear magnetic moments would precess at the same Larmor frequencies. In reality different nuclear moments precess at slightly different Larmor frequencies as a result of magnetic interactions between neighboring nuclei. The presence of microscopic magnetic fields created by the nuclei themselves or by electrons causes dephasing of nuclear spins in a sample. This results in an exponential decay of $\mathbf{M}_{tr}$ that occurs with a characteristic time constant T_2 known as *the spin–spin relaxation time* (Figure 1.3b). The dynamics of $\mathbf{M}_{tr}$ in the laboratory coordinate system is described by the following equations:

$$dM_x/dt = -M_x/T_2 + \gamma M_y B_0, \quad dM_y/dt = -M_y/T_2 - \gamma M_x B_0. \quad (1.3.3)$$

Equations (1.3.1) and (1.3.3) are known as *the Bloch equations*. By examining (1.3.3) we obtain the following solutions for M_x and M_y:

$$M_x = \exp(-t/T_2)[M_x(0)\cos(\omega_0 t) + M_y(0)\sin(\omega_0 t)],$$
$$M_y = \exp(-t/T_2)[-M_x(0)\sin(\omega_0 t) + M_y(0)\cos(\omega_0 t)]. \quad (1.3.4)$$

where $\omega_0 = \gamma B_0$, $M_x(0)$ and $M_y(0)$ are the initial values of the components of $\mathbf{M}_{tr}$. To express Eqs. (1.3.4) in a more compact form let

$$M_{xy} = M_x + jM_y, \quad (1.3.5)$$

where $j = \sqrt{-1}$. Using M_{xy} we can rewrite Eqs. (1.3.4) as

$$M_{xy} = M_{xy}(0)\exp(-j\omega_0 t - t/T_2), \quad (1.3.6)$$

where $M_{xy}(0)$ is the initial value of M_{xy}. This equation describes both T_2 decay and precession of the transverse magnetization about $\mathbf{B}_0$. It should be noted that the actual value of T_2 (as well as T_1) depends on the properties of nuclei and surrounding media, and the strength of $\mathbf{B}_0$. It can be shown that, in general, $T_2 \leq T_1$. The quantum-mechanical theory of nuclear relaxation is quite complicated and beyond the scope of this book (see [5,6] for an in-depth discussion of relaxation mechanisms and references to the relevant literature).

The Bloch equations can be modified in order to take into account the presence of an oscillating magnetic field $\mathbf{B}_1$. Everywhere in this book we will assume that $\mathbf{B}_1$ is much smaller than $\mathbf{B}_0$. Under this assumption the Bloch equations in the laboratory system can be written as follows:

$$\frac{d\mathbf{M}}{dt} = \gamma\mathbf{M} \times (\mathbf{k}B_0 + \mathbf{B}_1) - \mathbf{M}_{\mathrm{tr}}/T_2 - \mathbf{k}\frac{M_z - M_0}{T_1}. \qquad (1.3.7)$$

1.4. BASIC NMR EXPERIMENT

To avoid r.f. interference, detection of the NMR signal is typically performed after excitation of the transverse magnetization in a sample. According to the *principle of reciprocity* [7], the induced *emf* in the receiver r.f. coil is given by

$$emf = -\int \frac{\mathbf{B}_1}{I_c} \frac{\partial \mathbf{M}}{\partial t} \, dV. \qquad (1.4.1)$$

In this equation $\mathbf{B}_1/I_c$ is the magnetic field produced by a unit current in the coil at the location of $\mathbf{M}$. The integral in Eq. (1.4.1) is taken over the sample's volume. For simplicity we consider a long cylindrical coil of length L with N turns encompassing a sample of volume V. In this case $B_1/I_c = \mu_0(N/L)$, and the r.f. field is parallel to the axis of the coil. We further assume that the coil's axis is perpendicular to $\mathbf{B}_0$.

Based on the results obtained in the previous sections we can now describe a simple NMR experiment in which a system of nuclei is initially excited by a short 90 degree pulse with duration $t_{\mathrm{p}} \ll T_2, T_1$. Immediately after the pulse the nuclear magnetization lies in the transverse plane. Using Eqs. (1.3.4) and assuming that the Larmor frequency of nuclei $\omega_0 \gg 1/T_2$, we obtain

$$emf = \mu_0 N L^{-1} V \omega_0 M_0 e^{-t/T_2} \cos(\omega_0 t + \varphi), \qquad (1.4.2)$$

where the phase φ is dependent upon the excitation r.f. pulse. This equation shows that the induced signal, known as the *free induction decay* (FID), is proportional to $\omega_0 M_0$. Since $\omega_0 M_0 \propto n_I \gamma^3 I(I+1) B_0^2$, the sensitivity of signal detection in NMR experiments can be increased by using high field strengths and exciting nuclei that have large γ and are most abundant in the sample. Because ^{1}H nuclei satisfy the latter conditions in biological systems, they appear to provide the highest sensitivity in NMR studies *in vivo*. The subject of overall imaging sensitivity in terms of the signal-to-noise ratio is further discussed in Chapter 5.

NMR Spectra

The observed NMR signal can be conveniently analyzed via Fourier transformation of the signal. In practice the so-called *phase-sensitive detection* technique is used to shift signal down in frequency by ω_0 prior to analog-to-digital conversion (see Appendix). Assuming that signal acquisition starts at time $t = 0$, the NMR spectrum in the case of a single resonance can be written as

$$S(\omega) = \int_0^{\infty} K e^{-t/T_2 + j\varphi} e^{j\omega t} \, dt, \qquad (1.4.3)$$

where K is a real constant. Under the condition $e^{j\varphi} = 1$, the real $S_{\text{real}}(\omega)$ and $S_{\text{imag}}(\omega)$ components of the spectrum are given by:

$$S_{\text{real}}(\omega) = \frac{KT_2}{1 + T_2^2 \omega^2}, \quad S_{\text{imag}}(\omega) = \frac{KT_2^2 \omega}{1 + T_2^2 \omega^2}. \qquad (1.4.4)$$

The shape of $S_{\text{real}}(\omega)$ in (1.4.4) is known as *the Lorentzian lineshape*. The linewidth of the spectrum defined as the width of S_{real} at half height is $2/T_2$. In general, spectral linewidths as well as lineshapes also depend on other factors such as magnetic field heterogeneity and chemical exchange.

One of the main reasons responsible for the observed variety of NMR spectra is the chemical shift effect (CSE). CSE refers to the difference in Larmor frequencies of identical nuclei in different chemical environments. The basic phenomenon giving rise to CSE is the presence of microscopic currents induced by an external field $\mathbf{B}_0$. The induced currents in atoms and molecules shield nuclei from $\mathbf{B}_0$ by creating an additional magnetic field that is proportional to $\mathbf{B}_0$. As a result, the effective field acting on the nuclei can be expressed as

$$\mathbf{B} = \mathbf{B}_0(1 - \sigma), \qquad (1.4.5)$$

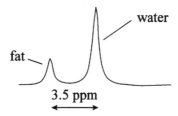

Figure 1.4. ^{1}H NMR spectrum obtained from a sample containing fat and water.

where σ is a constant (sometimes called the *shielding* or *screening constant*). It is a common practice to specify chemical shifts by using the dimensionless unit of parts per million (ppm). The chemical shift in ppm is defined as

$$\delta = \frac{\omega_i - \omega_r}{\omega_r} \times 10^6, \qquad (1.4.6)$$

where ω_i is the resonance frequency of the given nuclei and ω_r is an arbitrary chosen reference frequency. By using Eq. (1.4.5) and taking into account that normally $\sigma \ll 1$, we obtain

$$\delta = (\sigma_r - \sigma_i) \times 10^6, \qquad (1.4.7)$$

where σ_i and σ_r are the shielding constants of the given nuclei and of the reference nuclei, respectively. For example, due to stronger shielding ^{1}H nuclei in the fat tissue have a slightly lower Larmor frequency than that of ^{1}H nuclei in water. The observed fat/water chemical shift is approximately 3.5 ppm (Figure 1.4).

The chemical shift effect is important because it "fingerprints" chemically different populations of nuclei in a sample. That is, the presence of chemically shifted peaks in NMR spectra makes it possible to determine the chemical composition of a sample as well as to measure the relative amounts of nuclei present in different chemical environments.

1.5. T_2^* DECAY

Based on the Bloch equations, we could expect decay of the NMR signal to occur with a characteristic time T_2. However, in practice a more rapid FID, known as *the T_2^* decay*, is normally observed. To explain this faster decay we first notice that the local magnetic field in media is always heterogeneous because of the nonuniformity of an

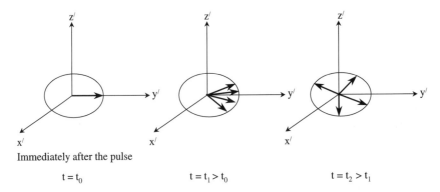

Figure 1.5. Dephasing of nuclear spins (indicated by arrows) in the rotating coordinate system after a 90 degree pulse.

applied field or spatially varying magnetic susceptibility in media. As a result, the Larmor frequency, ω_0, and phase of precession, $\varphi = \omega_0 t$, become spatially dependent. Second, we notice that in all NMR experiments the measured signal is actually a sum of signals produced by numerous nuclei in the excited volume. Since the spread of phases in the volume increases with time (Figure 1.5), the destructive interference of the signals produced by different nuclei causes faster decay of the net signal as compared with the T_2 decay.

To take into account the effect of dephasing due to macroscopic magnetic field inhomogeneities as well as T_2 decay caused by microscopic interactions between nuclei, the resultant decay of the transverse magnetization is frequently approximated by a phenomenological equation (compare with Eq. (1.3.6)):

$$M_{xy} = M_{xy}(0) \exp(-j\omega_0 t - t/T_2^*). \tag{1.5.1}$$

The rate of T_2^* decay can be written as

$$1/T_2^* = 1/T_2 + 1/T_2'. \tag{1.5.2}$$

In this equation T_2' is a time constant of the decay that occurs due to the presence of magnetic field inhomogeneity, δB. It should be noted that the exponential form of T_2^* decay can only be considered as an arbitrary assumption which makes it possible to conveniently describe the effect of magnetic field nonuniformity in certain cases.[1]

---

[1] An example of a nonexponential decay caused by magnetic field nonuniformity is presented in Chapter 6.

Magnetic field inhomogeneity, like chemical shift, is often expressed in terms of parts per million. The value of δB in ppm is defined as

$$\delta \tilde{B} = \frac{\delta B}{B_0} \times 10^6, \qquad (1.5.3)$$

where B_0 is an arbitrary chosen reference field. A rather crude estimate for T_2' is given by $(\gamma \delta B)^{-1}$. In the case when T_2' is short compared to T_2, the resultant signal decay is defined primarily by dephasing of spins. For example, $T_2' \approx 25\,\text{msec}$ for hydrogen nuclei at 1.5 Tesla and with local field inhomogeneities of 0.1 ppm. This T_2' value is significantly smaller than typical T_2 in biological tissues (40–150 msec) or T_2 for the ^{1}H nuclei in water (about 3 sec).

1.6. SPIN ECHOES

Fast dephasing of nuclear spins due to magnetic field heterogeneity causes a significant loss of signal and thereby presents a serious problem for NMR imaging and spectroscopy. With regard to NMR spectroscopy, another detrimental effect of dephasing is line broadening which impedes identification of closely spaced spectral lines. In practice magnetic field homogeneity is improved through the use of an auxiliary magnetic field that can be adjusted to reduce spatial variations in the resultant magnetic field in a sample (this approach known as *shimming* is further discussed in Chapter 9). However, it turns out that even with shimming residual magnetic field inhomogeneities remain. It is therefore extremely important that dephasing of spins can be significantly reduced by using the remarkable phenomenon discovered by E. Hahn and known as *spin echo* [8].

To describe this phenomenon, we consider a simple model system, composed of two noninteracting magnetic moments $\boldsymbol{\mu}_1$ and $\boldsymbol{\mu}_2$, in a static nonuniform magnetic field, $\mathbf{B} = \mathbf{k}B$. We can express $\mathbf{B}$ as a sum of a uniform field $\mathbf{B}_0$ and spatially varying field $\mathbf{B}'$

$$\mathbf{B} = \mathbf{k}(B_0 + B'). \qquad (1.6.1)$$

In the absence of r.f. irradiation, the dynamics of the magnetic moments in the rotating reference frame ($\omega = \gamma B_0$) is governed by the effective magnetic field $B - \omega/\gamma = B'$ (see Section 1.2). Suppose that the magnetic moments, initially aligned in the z direction, experience a 90 degree pulse such that immediately after the pulse $\boldsymbol{\mu}_1$ and $\boldsymbol{\mu}_2$ lie along the y'-axis (Figure 1.6(a)) in the rotating reference frame. Suppose also that a 180 degree pulse is subsequently applied along the x'-axis at time τ after the excitation. Figure 1.6(b) depicts $\boldsymbol{\mu}_1$ and

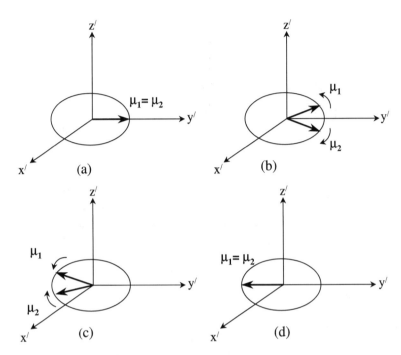

Figure 1.6. Refocusing of spins in the rotating coordinate system.

μ_2 immediately before the 180 degree pulse assuming that μ_1 and μ_2 have different Larmor frequencies $\omega_1 < \gamma B_0$ and $\omega_2 > \gamma B_0$, respectively. The 180 degree pulse inverts the y'-components of μ_1 and μ_2 (Figure 1.6(c)). By examining the diagram in Figure 1.6(c) we can see that at time 2τ (referred to as the *echo time*) the magnetic moments would be aligned again and would point in the negative y'-direction in the rotating reference frame (Figure 1.6(d)). Based on the fact that this result is valid for arbitrary moments μ_1 and μ_2, we can predict the phenomenon of refocusing of nuclear spins in a macroscopic sample that contains a great number of nuclei. The refocusing of spins manifests itself as a "spin-echo" signal (Figure 1.7).

To take into account the effect of internuclear interactions, we can consider the Larmor frequencies of spins as the sums of two components ω_{macro} and ω_{micro}, where ω_{macro} is defined by a macroscopic (static) magnetic field in media and ω_{micro} is defined by microscopic (time-dependent) magnetic fields produced by individual nuclei or by atomic electrons. Refocusing of spins following a 180 degree pulse requires that the Larmor frequencies of spins remain constant before

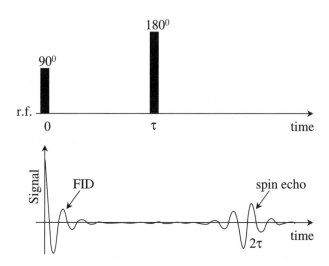

Figure 1.7. FID and spin-echo signals produced by a sequence of 90°–180° r.f. pulses.

and after the pulse. In reality these frequencies constantly change due to the presence of microscopic magnetic fields. The 180 degree pulse does not change the irreversible phase dispersion caused by temporal variations in ω_{micro}, and, therefore, does not alter T_2 decay of magnetization. In contrast, the phase dispersion caused by spatial variations in ω_{macro} is made zero at echo time 2τ. Consequently, spin echoes can be used for accurate measurement of T_2 relaxation time. For example, in the pulse sequence shown in Figure 1.7 the magnitude of the observed echo signal, S, can be written as

$$S(2\tau) = S(0)\exp(-2\tau/T_2), \qquad (1.6.2)$$

where $S(0)$ is the signal magnitude immediately after excitation. This equation can be used to calculate T_2 from the measured spin-echo signals acquired with different echo times 2τ.

It is a common practice to use a series of spin echoes created by a train of 180 degree pulses applied after a single excitation pulse. This increases the accuracy of T_2 measurements without lengthening the total measurement time. One of the most frequently used pulse sequences for T_2 measurements, known as the Carr–Purcell–Meiboom–Gill pulse sequence (CPMG), can be written using the notation $\{90^\circ_{x'} - \tau - 180^\circ_{y'} - 2\tau - 180^\circ_{y'} - 2\tau - 180^\circ_{y'}\ldots\}$. Notice that in the CPMG sequence the phase of the refocusing 180 degree pulses is shifted by $\pi/2$ with respect to the phase of the excitation pulse [9]. It can be shown that the CPMG sequence does not lead to accumulation of errors due to imperfections of 180 degree pulses in contrast to the Carr–Purcell sequence $\{90^\circ_{x'} - \tau - 180^\circ_{x'} - 2\tau - 180^\circ_{x'} - 2\tau - 180^\circ_{x'} \ldots\}$ [10], in which all r.f. pulses have the same phase.

It should be noted that Hahn has studied spin echoes produced by repetitive 90 degree pulses [8]. Hahn has found that in a sequence of two identical r.f. pulses separated by a time interval τ, a spin echo also occurs at time 2τ as in the case described earlier. Three identical r.f. pulses generate a more complicated pattern of spin echoes. Hahn has discovered that in the case when a third pulse is applied at time T after the first pulse, additional echoes occur at $T + \tau$, $2T - 2\tau$, $2T - \tau$, $2T$. Moreover, Hahn has also found that the amplitudes of the echoes generated by the third pulse, known as the *stimulated echoes*, in general depend on T_1 and/or T_2.

1.7. SIGNAL ATTENUATION DUE TO DIFFUSION

In his classic work devoted to spin echoes Hahn has shown that diffusion in a nonuniform magnetic field contributes to decay of the transverse magnetization in media [8]. To analyze the effect of diffusion we start with the equation that describes the dynamics of magnetization due to self-diffusion of particles in isotropic homogeneous media [6]:

$$\partial \mathbf{M}/\partial t = D\nabla^2 \mathbf{M}, \qquad (1.7.1.)$$

where D is the diffusion coefficient. Torrey [11] has demonstrated that in order to incorporate the effects of T_1 and T_2 relaxations and precession in an external magnetic field, the corresponding terms from the Bloch equations can be added to the right part of Eq. (1.7.1). Below we will obtain an exact solution of the Bloch equations in the case when diffusion of spins occurs in the presence of linear magnetic field gradients.

In the presence of a non-uniform magnetic field $\mathbf{B}$ given by a sum of constant and linear terms:

$$\mathbf{B} = \mathbf{k}(B_0 + \mathbf{Gr}), \qquad (1.7.2)$$

the dynamics of magnetization in the rotating reference frame is described by the modified Bloch equations:

$$\partial M_x/\partial t = -M_x/T_2 + \gamma \mathbf{Gr}M_y + D\nabla^2 M_x, \qquad (1.7.3a)$$

$$\partial M_y/\partial t = -M_y/T_2 - \gamma \mathbf{Gr}M_x + D\nabla^2 M_y. \qquad (1.7.3b)$$

Using $M_{xy} = M_x + jM_y$ again we obtain from the preceding two equations

$$\partial M_{xy}/\partial t = -M_{xy}/T_2 - j\gamma \mathbf{Gr}M_{xy} + D\nabla^2 M_{xy}. \qquad (1.7.4)$$

We seek a solution of this equation in the following form:

$$M_{xy} = M_{xy}(0)f(t) \exp\left(-t/T_2 - j\gamma\mathbf{r}\int_0^t \mathbf{G}\,dt'\right), \qquad (1.7.5)$$

where $f(t)$ is an arbitrary function of time. By substituting the last equation into Eq. (1.7.4) and taking into account that $f(0) = 1$ we obtain

$$f(t) = \exp\left(-D\gamma^2 \int_0^t \left[\int_0^{t'} \mathbf{G}\,dt''\right]^2 dt'\right). \qquad (1.7.6)$$

In a particular case when diffusion occurs in the presence of a time-independent gradient $\mathbf{G} = \mathbf{k}G_z$ we have $f(t) = \exp(-D\gamma^2 G_z^2 t^3/3)$. In this case M_{xy} is given by

$$M_{xy} = M_{xy}(0) \exp(-t/T_2 - j\gamma G_z zt - D\gamma^2 G_z^2 t^3/3). \qquad (1.7.7)$$

Equation (1.7.7) defines the attenuation of the transverse magnetization as a function of diffusion coefficient and gradient strength. In principle, this equation can be used to calculate the diffusion coefficient from a series of FIDs observed at different gradient strengths. However, in practice it can be difficult to differentiate between the effect of diffusion and signal attenuation due to dephasing of spins in the presence of an applied gradient. A better approach for NMR diffusion measurements was developed by Stejskal and Tanner [12]. The diffusion related attenuation of the NMR signal in the Stejskal–Tanner approach is achieved by applying two strong gradients symmetrically with respect to a 180 degree pulse (Figure 1.8). Because of the presence of a 180 degree pulse, the transverse magnetization of static spins is refocused at time TE after excitation. Therefore, the amplitude of the spin-echo signal from the static material is modulated by T_2 decay only. However, diffusion of spins in the direction of the applied gradients causes irreversible phase dispersion, which leads to additional signal attenuation. Under the assumption that the applied gradients are much greater than any intrinsic magnetic field gradients present in the sample, the effective attenuation of the

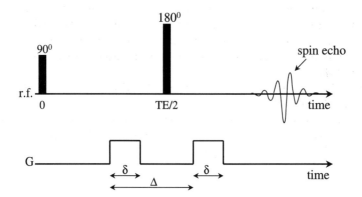

Figure 1.8. Stejskal–Tanner pulse sequence for diffusion measurements.

spin-echo signal is given by the well-known Stejskal–Tanner formula:

$$S(TE) = S(0) \exp[-TE/T_2 - D\gamma^2 G^2 \delta^2 (\Delta - \delta/3)], \qquad (1.7.8)$$

where $S(0)$ is the signal magnitude immediately after the excitation pulse; G is the amplitude of the applied gradients; δ and Δ are the duration and time interval between the pulsed gradients, respectively (Figure 1.8). Measurements of the diffusion coefficient in the Stejskal–Tanner method can be carried out by varying the area under the gradient pulse, $G\delta$, or by varying Δ.

REFERENCES

[1] F. Bloch, W.W. Hansen, M. Packard. "Nuclear induction," *Phys. Rev.* **69**, 127 (1946).

[2] F. Bloch, W.W. Hansen, M. Packard. "The nuclear induction experiment," *Phys. Rev.* **70**, 474 (1946).

[3] F. Bloch. "Nuclear induction," *Phys. Rev.* **70**, 460 (1946).

[4] E.M. Purcell, H.C. Torrey, R.V. Pound. "Resonance absorption by nuclear magnetic moments in a solid," *Phys. Rev.* **69**, 37 (1946).

[5] A. Abragam. *Principles of Nuclear Magnetism.* Oxford University Press (1983).

[6] C.P. Slichter. *Principles of Magnetic Resonance.* Springer-Verlag (1978).

[7] D.I. Hoult, R.E. Richards. "The signal-to-noise ratio of the nuclear magnetic resonance experiment," *J. Magn. Reson.* **24**, 71 (1976).

[8] E.L. Hahn. "Spin echoes," *Phys. Rev.* **80**, 580 (1950).

[9] S. Meiboom, D. Gill. "Modified spin-echo method for measuring nuclear relaxation times," *Rev. Sci. Instr.* **29**, 688 (1958).

[10] H.Y. Carr, E.M. Purcell. "Effects of diffusion on free precession in nuclear magnetic resonance experiments," *Phys. Rev.* **94**, 630 (1954).

[11] H.C. Torrey. "Bloch equations with diffusion terms," *Phys. Rev.* **104**, 563 (1956).

[12] E.O. Stejskal, J.E. Tanner. "Spin diffusion measurements: spin echoes in the presence of a time-dependent field gradient," *J. Chem. Phys.* **42**, 288 (1965).

CHAPTER 2

Excitation of the Transverse Magnetization

In NMR imaging and spectroscopy it is a common practice to excite repeatedly a macroscopic system of nuclei by a series of r.f. pulses. For example, in the conventional approach used for spatial encoding in MRI (see Chapter 3) a number of excitations (e.g., 128 or 256) are needed to collect all spatially encoded signals required for subsequent image reconstruction. One of the main assumptions of this approach is the existence of a steady-state, under which repetitive excitations would produce the same transverse magnetization in a sample. Using the Bloch equations we demonstrate that such a steady state can indeed be established as a result of the evolution of the nuclear magnetization subjected to a series of r.f. pulses. The last section of this chapter describes an example of spatially selective excitation that creates nonzero transverse magnetization only in a chosen slice of material.

2.1. DYNAMICS OF REPEATEDLY EXCITED MAGNETIZATION

In this section we consider the dynamics of nuclear magnetization in an external magnetic field $\mathbf{B}_0$ in the presence of a train of identical r.f. pulses (Figure 2.1(a)). To simplify further derivations we assume that the transverse magnetization in a sample is negligibly small immediately before the beginning of each successive excitation. This assumption is justified when the sequence repetition time, TR, is much longer than T_2 or when external magnetic field gradients

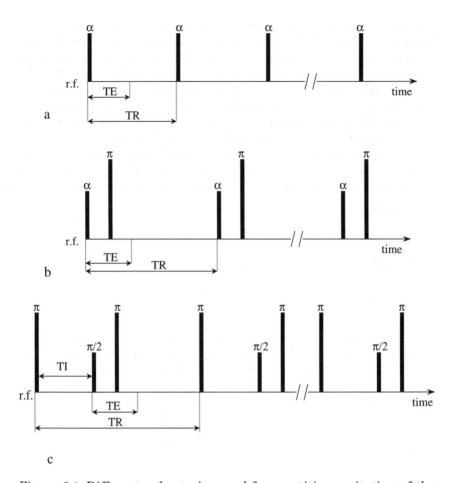

Figure 2.1. Different pulse trains used for repetitive excitation of the transverse magnetization.

(sometimes called *crusher* or *spoiling gradients*) are applied after signal acquisition in order to disperse the transverse magnetization in a sample. A general case, when the steady-state transverse magnetization is nonzero immediately before the beginning of an r.f. pulse, is considered in a classic paper by Ernst and Anderson [1].

During an excitation the effective magnetic field in the rotating reference frame is given by (see Chapter 1):

$$B_{\text{eff}} = [(\omega_0 - \omega_L)^2/\gamma^2 + B_{\text{r.f.}}^2]^{1/2}, \tag{2.1.1}$$

where ω_0 is the frequency of the r.f. field and ω_L is the local Larmor frequency. Let $\omega_L = \gamma B_0 + \delta\omega$, where B_0 is uniform and $\delta\omega$ is defined by magnetic field nonuniformity or chemical shift effect. In the following derivations we assume that the irradiating r.f field is applied at the resonant frequency (i.e., $\omega_0 = \gamma B_0$). We also assume that the amplitude of the r.f. field is large enough such that $\delta\omega \ll \gamma B_{\text{r.f.}}$. Under the latter condition the effective magnetic field in Eq. (2.1.1) is equal to $B_{\text{r.f.}}$ and is independent of the local Larmor frequency of nuclei.

Using the Bloch equations (see Chapter 1) and assuming that the duration of the excitation pulses is shorter than T_1 and T_2, we obtain the following equations for the longitudinal magnetization M_z:

$$M_z(0) = M_0,$$

$$M_z(TR) = M_0[(1 - e^{-TR/T_1}) + e^{-TR/T_1}\cos\alpha],$$

$$M_z(2TR) = M_0[(1 - e^{-TR/T_1})(1 + e^{-TR/T_1}\cos\alpha) + e^{-2TR/T_1}\cos^2\alpha],$$

$$M_z(nTR) = M_0\left[(1 - e^{-TR/T_1})\sum_{k=0}^{n-1}e^{-kTR/T_1}\cos^k\alpha + e^{-nTR/T_1}\cos^n\alpha\right].$$

In the preceding equations the z-axis is chosen in the direction of $\mathbf{B}_0$; $M_z(nTR)$ is the longitudinal magnetization immediately before the $(n+1)$ pulse; α is the angle of rotation of the magnetization (known as the *flip angle*) in the rotating reference frame (see Chapter 1); M_0 is the equilibrium magnetization. Using the formula for geometric progression the expression for $M_z(nTR)$ can be written as

$$M_z(nTR) = M_0\left[(1 - e^{-TR/T_1})\frac{1 - e^{-nTR/T_1}\cos^n\alpha}{1 - e^{-TR/T_1}\cos\alpha} + e^{-nTR/T_1}\cos^n\alpha\right].$$

$$(2.1.2)$$

This equation describes the longitudinal magnetization as a function of T_1 relaxation time, repetition time, flip angle, and number of excitations. As the number of excitations increases, the longitudinal magnetization reaches its steady-state

$$M_z(TR) = M_0\frac{1 - E_1}{1 - E_1\cos\alpha}, \qquad (2.1.3)$$

where $E_1 = e^{-TR/T_1}$. Note that in the case when $\alpha = \pi/2$, the steady-state is established after the first pulse regardless of the ratio TR/T_1. In a general case, however, the relaxation to the steady-state in (2.1.3) occurs much more slowly and the relaxation rate depends upon α as well as TR/T_1 (Figure 2.2).

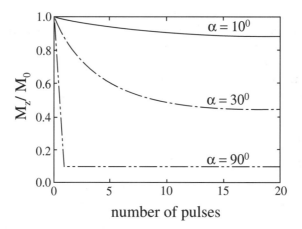

Figure 2.2. Dependence of the longitudinal magnetization $M_z(nTR)$ on the number of excitations at T_1 of 800 ms and TR of 80 ms.

Consider the equation for the steady-state transverse magnetization at an arbitrary time TE after an excitation. In the case when dephasing of spins can be neglected (i.e., when $\delta\omega TE \ll 1$), the steady-state transverse magnetization in the rotating reference frame can be expressed as $M_{tr}(TE) = M_z(TR)\, e^{-TE/T_2} \sin\alpha$. Using (2.1.3) we obtain

$$M_{tr}(TE) = M_0 E_2 \sin\alpha \frac{1 - E_1}{1 - E_1 \cos\alpha}, \qquad (2.1.4)$$

where $E_2 = e^{-TE/T_2}$. Using this equation it is easy to verify that the steady-state transverse magnetization decreases with decreasing TR. This conclusion also follows from the observation that short TR does not allow enough time for relaxation of the longitudinal magnetization in a sample after an excitation pulse. In this case the sample is frequently referred to as *saturated*. Equation (2.1.4) shows that M_{tr} increases with decreasing TE because of decreased T_2 decay. These simple results turn out to be extremely important because they can be used to enhance contrast in MR images (see Chapters 4 and 8). The dependence of M_{tr} on flip angle is discussed in section 2.2.

Spin-Echo Sequence

To avoid signal loss due to magnetic field nonuniformity, 180 degree pulses are frequently included in pulse sequences used in NMR imaging and spectroscopy. A spin-echo pulse sequence is shown in Figure 2.1(b). The dynamics of magnetization in this case can be studied by utilizing the same approach we used to derive Eq. (2.1.3). However, to avoid cumbersome derivations we will consider at this time only the equations describing the steady-state magnetization.

Let $M_{z-}(TR + TE/2)$ and $M_{z+}(TR + TE/2)$ denote the steady-state longitudinal magnetization immediately before and after a 180 degree pulse, respectively. From the Bloch equations it follows that

$$M_{z-}(TR + TE/2) = [M_z(TR)\cos\alpha - M_0]\, e^{-TE/2T_1} + M_0,$$

$$M_{z+}(TR + TE/2) = -M_{z-}(TR + TE/2),$$

$$M_z(TR) = [M_{z+}(TR + TE/2) - M_0]\, e^{-(TR - TE/2)/T_1} + M_0,$$

where $M_z(TR)$ is the steady-state magnetization immediately before an excitation pulse. Using these equations we obtain

$$M_z(TR) = M_0 \frac{1 + e^{-TR/T_1} - 2e^{-(TR - TE/2)/T_1}}{1 + e^{-TR/T_1}\cos\alpha}. \qquad (2.1.5)$$

From (2.1.5) it follows that the steady-state transverse magnetization at time TE after excitation is given by

$$M_{tr}(TR) = M_0\, e^{-TE/T_2}\sin\alpha \frac{1 + e^{-TR/T_1} - 2e^{-(TR - TE/2)/T_1}}{1 + e^{-TR/T_1}\cos\alpha}. \qquad (2.1.6)$$

Inversion-Recovery Sequence

A basic inversion-recovery pulse sequence is shown in Figure 2.1(c). During each cycle the first 180 degree pulse (referred to below as the *inversion pulse*) inverts the longitudinal magnetization, which then recovers during the time interval *TI*, known as the *inversion time*. The subsequent excitation and refocusing of the magnetization are essentially the same as in a spin-echo sequence (Figure 2.1(b)). It is clear that M_z is zero immediately after a 90 degree pulse. Following the pulse the longitudinal magnetization recovers during time $TE/2$. Taking into account that M_z is subsequently inverted again by a 180 degree pulse applied at time $TE/2$ after the 90 degree pulse we obtain:

$$M_z(TR) = M_0[1 - 2e^{-(TR - TI - TE/2)/T_1} + e^{-(TR - TI)/T_1}], \qquad (2.1.7)$$

where $M_z(TR)$ denotes the steady-state longitudinal magnetization immediately before an inversion pulse.

The steady-state transverse magnetization at time TE after a 90 degree pulse (Figure 2.1(c)) can be expressed as

$$M_{tr}(TE) = [M_0 - (M_z(TR) + M_0)\, e^{-TI/T_1}]\, e^{-TE/T_2}$$

$$= M_0 e^{-TE/T_2}[1 - 2e^{-TI/T_1} + 2e^{-(TR - TE/2)/T_1} - e^{-TR/T_1}]. \qquad (2.1.8)$$

In the case when $TE \ll T_1$ we can simplify this equation as follows:

$$M_{tr}(TE) = M_0\, e^{-TE/T_2}[1 - 2e^{-TI/T_1} + e^{-TR/T_1}]. \qquad (2.1.9)$$

According to Eq. (2.1.9), it is possible to null the signal from specific nuclei with known T_1 value by choosing the inversion time to be

$$TI = T_1 \ln \left[\frac{2}{1 + e^{-TR/T_1}} \right]. \qquad (2.1.10)$$

This feature of inversion-recovery sequence is often used in diagnostic MRI in order to eliminate the signal from a chosen tissue such as fat or cerebrospinal fluid (CSF). For example when a lesion in the brain parenchyma is bright on a T_2-weighted image (see Chapter 4) but obscured by the even brighter CSF in the adjacent ventricle, suppression of the CSF signal through the use of inversion-recovery sequence can improve the contrast between the brain tumor and the surrounding tissue.

2.2. ERNST ANGLE

Steady-state transverse magnetization established in the presence of r.f. pulses depends on a number of parameters, which can be divided into two different groups. The first group is composed of parameters such as M_0, T_1, and T_2 relaxation times that are defined by the properties of the excited nuclei, their temperature and surroundings, and the strength of the applied magnetic field. The second group consists of user-controlled parameters, such as TR, TE, and flip angle, which can be chosen to maximize the observed signal or enhance image contrast (see Chapter 4). It is easy to verify that the maximum transverse magnetization in (2.1.4) and (2.1.6) is achieved when $TR \gg T_1$ and $\alpha = \pi/2$. However, in order to ensure desirable image contrast or shorten scan time (see Chapters 4 and 7) it is often necessary to use TR shorter than T_1. In the latter case Eqs. (2.1.4) or (2.1.6) can be used to determine the optimum flip angle, known as the *Ernst angle*, which corresponds to the maximum M_{tr} (Figure 2.3).

Let us first consider Eq. (2.1.4) which describes the steady-state transverse magnetization in the pulse sequence in Figure 2.1(a). Using Eq. (2.1.4) it is easy to show that the Ernst angle, α_E, satisfies the following equation [1]

$$\cos \alpha_E = \exp(-TR/T_1). \qquad (2.2.1)$$

From Eq. (2.2.1) it follows that α_E increases with TR. In the case when $TR \gg T_1$ we have $\alpha_E \approx \pi/2$ and $M_{tr}(TE) \approx M_0 \exp(-TE/T_2)$.

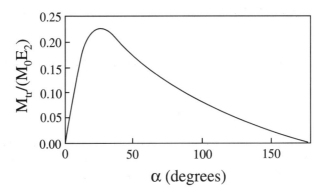

Figure 2.3 Dependence of the transverse magnetization on flip angle at $TR/T_1 = 0.1$.

Conversely, in the case when $TR \ll T_1$ the Ernst angle is approximately equal to $(2TR/T_1)^{1/2}$ and $M_{tr}(TE) \approx M_0(TR/2T_1)^{1/2} \exp(-TE/T_2)$.

The choice of flip angle becomes particularly important in the case of rapid gradient-echo MR imaging (see Chapter 7) which is performed with short TR to decrease scan time. Note that when imaging with $TR \ll T_1$ and large flip angle (i.e. $\alpha \approx 90$ degrees) the signal is proportional to TR/T_1. Conversely, when imaging with the Ernst flip angle, the signal varies as $(TR/T_1)^{1/2}$. As a result, to avoid a significant loss of signal, gradient-echo imaging with short TR is typically implemented with low flip angle that does not significantly deviate from the Ernst angle.

From Eq. (2.1.6) it follows that the Ernst angle for a spin-echo sequence is defined by the equation

$$\cos \alpha_E = -\exp(-TR/T_1). \qquad (2.2.2)$$

Note that, in contrast to the previous case, the Ernst angle for a spin-echo sequence is greater than $\pi/2$.

2.3. SPATIALLY SELECTIVE EXCITATION

Magnetic resonance imaging is typically used to image an object in three dimensions. To achieve spatial localization it is a common practice to first selectively excite transverse magnetization in a thin slice of material, which is subsequently imaged in the two remaining directions. The design of spatially selective pulses is an interesting and important subject that has been studied by many investigators. To elucidate the principles of spatially selective excitation, we will consider the dynamics of the transverse magnetization in the presence of a *sinc* shaped r.f. pulse. The approach used in this section is based on the small-flip-angle approximation that makes it possible to solve the Bloch equations analytically [2–4].

Suppose that a short r.f. pulse with duration $T_p \ll T_2, T_1$ is applied in the presence of an external magnetic field with linear gradient: $\mathbf{B} = \mathbf{k}(B_0 + Gz)$, where $\mathbf{k}$ is a unit vector in the z direction. During the excitation the dynamics of the nuclear magnetization in the rotating reference frame is described by the following equations:

$$\frac{dM_x}{dt} = \gamma G z M_y,$$

$$\frac{dM_y}{dt} = -\gamma G z M_x + \gamma B_{\text{r.f.}}(t) M_z,$$

$$\frac{dM_z}{dt} = -\gamma B_{\text{r.f.}}(t) M_y,$$

assuming that the r.f. field has only x-component, $B_{\text{r.f.}}(t)$.

The small-flip-angle approximation is based on the assumption that during excitation the magnetization remains close to the z-axis so that $(M_x^2 + M_y^2)^{1/2} \ll M_z \approx M_0$. Let

$$M_{xy} = M_x + jM_y. \tag{2.3.1}$$

Neglecting the difference between M_z and M_0 we obtain:

$$\frac{dM_{xy}}{dt} = -j\gamma G z M_{xy} + j\gamma B_{\text{r.f.}}(t) M_0. \tag{2.3.2}$$

The solution of the last equation with initial condition $M_{xy}(0) = 0$ is given by

$$M_{xy} = j\gamma M_0 \int\limits_0^t B_{\text{r.f.}}(s) \exp\left[j\gamma z \int\limits_t^s G(u)\, du \right] ds. \tag{2.3.3}$$

This equation defines the transverse magnetization as a function of the time-varying r.f. field and applied gradient. In principle, Eq. (2.3.3) can be used to design r.f. pulses and gradient waveforms for excitation of arbitrary spatial profiles of the transverse magnetization in a sample. Several interesting examples of spatially selective excitation in one and two dimensions have been described by Pauly *et al.* [4].

An important practical example is the excitation by a *sinc*-shaped r.f. pulse

$$B_{\text{r.f.}}(t) = A \frac{\sin \omega_b(t - t_0)}{\omega_b(t - t_0)}, \tag{2.3.4}$$

where A and ω_b are constants. The spectrum of the pulse in (2.3.4) is given by

$$\hat{B}_{\mathrm{r.f.}}(\omega) = \begin{cases} const \cdot e^{j\omega t_0}, & \text{if } |\omega| < \omega_b \\ 0, & \text{if } |\omega| > \omega_b \end{cases} \qquad (2.3.5)$$

assuming that $\omega_b > 0$. A *sinc*-shaped pulse with the spectrum defined in Eq. (2.3.5) will excite only the nuclei with Larmor frequencies centered near the irradiating frequency $\omega_0 = \gamma B_0$ within the bandwidth of the pulse, $2\omega_b$.[1] In the presence of a static, linear gradient, Larmor frequency of nuclei becomes spatially dependent: $\omega_L = \omega_0 + \gamma G z$. Because of the spatially varying ω_L, excitation of the transverse magnetization by a *sinc*-shaped pulse occurs only inside a slice of material that contains nuclei with Larmor frequencies $|\omega_L - \omega_0| \leq \omega_b$. It then follows that the excited slice thickness is given by

$$\delta z = 2|\omega_b/\gamma G|. \qquad (2.3.6)$$

We can use Eq. (2.3.3) to determine the transverse magnetization generated by a *sinc* pulse. Under the condition that G remains constant during the pulse, this equation can be written as

$$M_{xy}(T_p) = j\gamma A M_0 \exp\left(-\frac{j\gamma G T_p z}{2} \right) \int_{-T_p/2}^{T_p/2} \exp(j\gamma G z u) \frac{\sin \omega_b u}{\omega_b u} \, du, \qquad (2.3.7)$$

where $T_p = 2t_0$ is the duration of the pulse. In the case when $T_p \gg 1/\omega_b$, the integral in Eq. (2.3.7) can be approximated by Fourier transform of the *sinc* function and the resulting expression for M_{xy} is given by

$$M_{xy}(T_p) = \begin{cases} \dfrac{j\pi\gamma A M_0}{\omega_b} e^{-j\gamma G T_p z/2}, & \text{if } |\gamma G z| < \omega_b \\ 0, & \text{if } |\gamma G z| > \omega_b \end{cases} \qquad (2.3.8)$$

Notice that according to the above equation the transverse magnetization acquires a phase $\varphi = -\gamma G T_p z/2$. Because of the accumulated phase, the signal from the excited slice will be very small if no measures to cancel this phase are taken. Fortunately, refocusing of

---

[1] A *sinc*-shaped r.f. pulse applied in the absence of external field gradients provides an example of a *frequency (spectrally) selective pulse* that only excites a limited range of frequencies. In MRI frequency selective pulses are often used for chemical shift imaging (see Chapter 6).

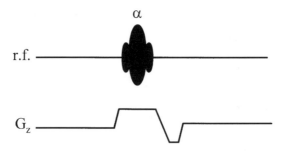

Figure 2.4. Schematic of spatially selective excitation. After the excitation r.f. pulse, the polarity of the slice-select gradient is reversed in order to refocus the transverse magnetization in the excited slice.

the magnetization can be easily achieved by reversing the polarity of the slice-select gradient after the excitation (Figure 2.4). It is easy to verify that the area of the refocusing gradient lobe must be half of the area of the gradient lobe used for slice selection in order to null the phase accumulated during excitation. Figure 2.5 shows an example of a 25 degree *sinc* pulse applied along the x-axis in the rotating frame of reference and the corresponding components of the transverse magnetization calculated numerically. The deviations from the ideal excitation profile are mostly due to a small number of side lobes of the *sinc* function used in this example. Although the slice profile can be improved by using a larger number of side lobes, the penalty is prolonged pulse duration which is undesirable for a number of applications such as flow imaging, rapid gradient-echo imaging etc., when short excitation pulses (a few milliseconds) are needed to reduce dephasing of spins or shorten repetition time.

The small-flip-angle approximation makes it possible to design r.f. pulses with flip angles on the order of $\pi/2$. Spatially selective pulses with flip angles greater than $\pi/2$ are frequently designed by using numerical solutions of the Bloch equations. Alternatively, r.f. pulses can be designed by using an approach independently suggested by Shinnar and Le Roux [5].

An important problem in the design of spatially selective pulses is the deviation from the intended excitation profile caused by the heterogeneity of the r.f. field. Distortion of slice profile can be particularly severe in the case when r.f. pulses are generated by using surface r.f. coils which produce an extremely inhomogeneous $B_{\mathrm{r.f.}}$ field. The effect of r.f. inhomogeneity can be reduced by using so called "adiabatic pulses" [6,7] which are designed to compensate for the spatial variations

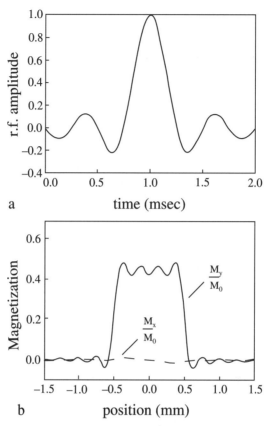

Figure 2.5. Example of spatially selective excitation: (a) amplitude of a 25 degree *sinc* pulse applied along the x axis in the rotating reference frame; (b) transverse components of the magnetization in the rotating reference frame after the excitation and gradient refocusing.

in $B_{r.f.}$ and are characterized by simultaneous modulation of the r.f. amplitude and frequency.

REFERENCES

[1] R.R. Ernst, W.A. Anderson. "Application of Fourier transform spectroscopy to magnetic resonance," *Rev. Scien. Instr.* **37**, 93 (1965).

[2] D.I. Hoult. "The solution of the Bloch equations in the presence of a varying B_1 field – an approach to selective pulse analysis," *J. Magn. Reson.* **35**, 69 (1979).

[3] W.S. Hinshaw, A.H. Lent. "An introduction to NMR imaging: from the Bloch equations to the imaging equation." *Proc. IEEE* **71**, 338 (1983).

[4] J. Pauly, D. Nishimura, A. Macovski. "A k-space analysis of small-tip-angle excitation." *J. Magn. Reson.* **81**, 43 (1989).

[5] J. Pauly, P. Le Roux, D. Nishimura, A. Macovski. "Parameter relations for the Shinnar–Le Roux selective excitation pulse design algorithm." *IEEE Trans. Med. Imag.* **10**, 53 (1991).

[6] M.S. Silver, R.I. Joseph, D.I. Hoult. "Selective spin inversion in nuclear magnetic resonance and coherent optics through an exact solution of the Bloch–Riccati equation." *Phys. Rev.* **A 31**, 2753 (1985).

[7] J. Baum, R. Tycko, A. Pines. "Broadband and adiabatic inversion of a two-level system by phase-modulated pulses." *Phys. Rev.* **32**, 3435 (1985).

CHAPTER 3

Basic Techniques for 2D and 3D MRI

3.1. IMAGE RECONSTRUCTION FROM DISCRETE SAMPLES

Because image reconstruction in MRI typically involves the use of Fourier transform (FT), we need to begin our discussion by describing the basic principles of Fourier analysis. The continuous Fourier transform of a function $f(x)$ is defined as

$$S(k_x) = \int_{-\infty}^{\infty} f(x)e^{jk_x x}\, dx, \qquad (3.1.1)$$

where $j = \sqrt{-1}$. If the Fourier transform $S(k_x)$ is known, then the function $f(x)$ is given by inverse Fourier transform of $S(k_x)$:

$$f(x) = \frac{1}{2\pi} \int_{-\infty}^{\infty} S(k_x)e^{-jk_x x} dk_x. \qquad (3.1.2)$$

For the purpose of convenience, we will refer to $f(x)$ as the object. The process of measuring $S(k_x)$ will be referred to as sampling, and the function $S(k_x)$ itself will be referred to as the signal.

In practice it is not possible to obtain $S(k_x)$ as a function of continuous k_x. Instead $S(k_x)$ is measured only at a finite number of sampling points. To demonstrate how discrete samples of the signal can be used for image reconstruction, let us assume that $S(k_x)$ in Eq. (3.1.1) is measured at N locations:

$$k_x = 2\pi n/L_x. \qquad (3.1.3)$$

In this equation n is an integer, $-N/2 \leq n < N/2$, and N is assumed to be even; L_x is a positive constant, known as *field-of-view*, which defines the size of the imaged region. Using Equations (3.1.1) and (3.1.3) we obtain

$$S(n) \equiv S(k_x(n)) = \int_{-\infty}^{\infty} f(x)e^{j2\pi nx/L_x}\, dx. \qquad (3.1.4)$$

We now define the reconstructed image intensity, I, as the inverse discrete FT of the signal in (3.1.4) [1]:

$$I(p) = (1/N) \sum_{n=-N/2}^{n=N/2-1} S(n)\exp(-j2\pi np/N). \qquad (3.1.5)$$

In this equation p is an integer, $-N/2 \leq p < N/2$.[1] After substituting $S(k_x)$ from Eq. (3.1.4) into Eq. (3.1.5) and using the formula for geometric progression, we obtain the following equation describing the relationship between the object and its image:

$$I(p) = \int_{-\infty}^{\infty} f(x)PSF(x - x_p)\, dx, \qquad (3.1.6)$$

where $x_p = pL_x/N$ and

$$PSF(x) = \frac{e^{-j\pi x/L_x}\sin(\pi Nx/L_x)}{N\sin(\pi x/L_x)}. \qquad (3.1.7)$$

From Eq. (3.1.6) it follows that the image intensity is given by the convolution of the object $f(x)$ and $PSF(x)$, known as *point-spread function*. The term point-spread function is used because $PSF(x)$ represents the image of a point source described by the Dirac delta function, $\delta(x)$.[2]

These results can be generalized to describe Fourier reconstruction from discrete samples in two or three dimensions. For example, a two-dimensional image intensity is given by the convolution of

---

[1] Discrete FT of the signal can be computed very efficiently by using the approach described by Cooley and Tukey in 1965 and known as the *fast Fourier transform* (FFT). The majority of FFT algorithms require that the number of data points be an integer power of two. As a result, the number of signal samples in applications which entail the use of FFT is most likely to be a power of two.

[2] The Dirac delta function has the properties:

$$\text{if } x \neq 0 \quad \text{then} \quad \delta(x) = 0$$

and

$$\int_{-\infty}^{+\infty} \delta(x)\, dx = 1.$$

the object and a two-dimensional point-spread function $PSF(x,y)$, which is given by the product of one-dimensional point-spread functions $PSF_x(x)$ and $PSF_y(y)$.

Spatial resolution

Image intensity in Eq. (3.1.6) can be expressed as a sum of weighted elements $f(x)\,dx$ with $PSF(x)$ as the weighting function [2]. It is therefore clear that the image intensity has contributions from different locations in the object. The ability of an image to accurately reproduce spatial variations in the object is characterized by the spatial resolution of the image. Spatial resolution is often defined as the minimum distance between two points in an object at which they can still be distinguished from one another in the image. Based on Eq. (3.1.6) spatial resolution, δx, can also be defined as the effective width of a point-spread function [3]:

$$\delta x = \frac{1}{PSF(0)} \int_{-L_x/2}^{L_x/2} PSF(x)\,dx. \tag{3.1.8}$$

By using the latter definition and Eq. (3.1.7) it can be shown that $\delta x = L_x/N$.

Aliasing

A one-dimensional point-spread from Eq. (3.1.7) is plotted in Figure 3.1. Notice that the $PSF(x)$ oscillates such that its amplitude reaches maximum at the origin $x = 0$ and rapidly decreases with increasing distance from the origin. Since $PSF(x)$ is also a periodical function with period L_x, the reconstructed image intensity $I(p)$ is defined primarily by a sum of the equally weighted object intensities at $x = \frac{pL_x}{N} + mL_x$, where $m = 0, \pm 1, \pm 2 \ldots$. This phenomenon (known as *aliasing*) results from discrete sampling of $S(k)$ and in general prevents complete recovery of $f(x)$. However, under the circumstances typical for MR imaging the object has a limited spatial extent such that

$$\begin{cases} |f(x)| \geq 0, & \text{if } -a/2 \leq x \leq a/2 \\ |f(x)| = 0, & \text{otherwise.} \end{cases} \tag{3.1.9}$$

According to the sampling theorem [1], in this case aliasing can be avoided and $f(x)$ in principle can be fully recovered by satisfying the Nyquist criterion for the sampling interval Δk:

$$\Delta k \leq \frac{2\pi}{a}. \tag{3.1.10}$$

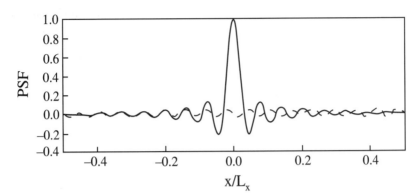

Figure 3.1. One-dimensional point-spread function with $N = 32$. The solid and dashed lines represent the real and imaginary parts of the *PSF*, respectively.

By examining this equation and Eq. (3.1.3) we conclude that in order to prevent aliasing the selected field-of-view L_x should be equal to or greater than the length of the object, a.

3.2. FREQUENCY AND PHASE ENCODING

In practice MR imaging is most frequently performed by using the approach suggested by Kumar *et al.* [4]. This approach employs a sequence of pulsed magnetic field gradients which are applied during a free induction decay or a spin echo to ensure that the signal is given by Fourier transform of the transverse magnetization in the sample. At this stage we are ready to discuss two basic imaging techniques for Fourier encoding: *frequency encoding* and *phase encoding*.

Frequency Encoding
Frequency encoding is implemented by acquiring signal in the presence of an external magnetic field gradient. The purpose of the gradient, known as the *frequency-encoding* or *readout gradient*, is to make the Larmor frequency of nuclei spatially dependent during signal acquisition.

The signal acquired in the presence of a readout gradient is composed of components with frequencies from a narrow range, known as the *signal bandwidth*, around the Larmor frequency ω_0 at the center of the field-of-view. The frequency ω_0 is also known as the *reference frequency*. Each of the signal components is produced by spins from a certain location in the object. Because the typical signal bandwidth is much smaller than ω_0,[3] it is important to remove the reference frequency from the signal in order to distinguish between the spectral components within the bandwidth. In order to do this a special technique, referred to as the *phase-sensitive detection* (see Appendix), is used to shift the signal down in frequency by ω_0. As a result the output signal consists of primarily low frequency components.[4]

At this stage we consider signal acquisition in the presence of a readout gradient, $G_x = dB_z/dx$, given by

$$G_x = \begin{cases} -G_{0,x}, & t_0 \leq t \leq t_0 + T_s/2 \\ G_{0,x}, & t_0 + T_s/2 < t \leq t_0 + 3T_s/2 \\ 0, & \text{elsewhere} \end{cases} \qquad (3.2.1)$$

where $G_{0,x}$ is a constant (Figure 3.2). Suppose that the signal acquisition starts at $t_0 + T_s/2$ and ends at $t_0 + 3T_s/2$. The phase of the transverse magnetization accumulated during the interval $[t_0, t_0 + T_s/2]$ in the presence of the first (defocusing) lobe of the readout gradient can be written as

$$\phi_0(x) = \gamma x \int_{t_0}^{t_0 + T_s/2} G_x \, dt = -\gamma G_{0,x} T_s x/2. \qquad (3.2.2)$$

The readout gradient reversal at $t = t_0 + T_s/2$ initially causes refocusing of spins, which in turn leads to a signal increase, known as the *gradient echo*, such that the maximum signal occurs in the center of the acquisition interval (Figure 3.2). It can easily be shown that during the acquisition the phase of the transverse magnetization is given by

$$\phi(x,t) = \gamma G_{0,x}(t - t_0)x - \gamma G_{0,x} T_s x. \qquad (3.2.3)$$

Suppose that N samples of the signal are collected at $t = t_0 + T_s/2 + p\tau_w$, where an integer p changes from 0 to $N - 1$. The sampling interval $\tau_w = T_s/N$ is known as the *dwell time* and T_s is referred to as the *acquisition* or *readout time*. Using Eq. (3.2.3) we can express the phase of magnetization as

$$\phi(x, t(n)) = \gamma G_{0,x} \tau_w n x, \qquad (3.2.4)$$

where $n = p - N/2$ and $t(n) = t_0 + T_s + n\tau_w$.

---

[3] For example, the Larmor frequency of ^{1}H nuclei at 1.5 Tesla is about 63.86 MHz, whereas typical signal bandwidth in MRI is less than 100 Khz.
[4] Strictly speaking, during phase-sensitive detection the signal is shifted in frequency by $+/-\omega_0$. Therefore, the output signal also includes high-frequency components that are centered at $2\omega_0$. However, these components are subsequently removed from the signal by passing them through a low-pass filter.

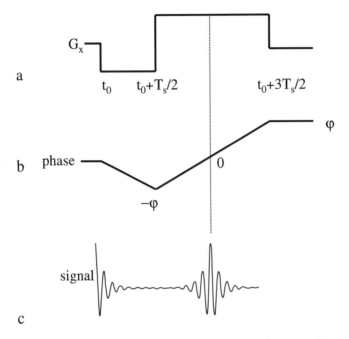

Figure 3.2. Formation of a gradient-echo: (a) readout gradient wave-form; (b) accumulated phase; (c) NMR signal.

To understand how NMR signal obtained with frequency encoding can be used for image reconstruction we need to consider the equation describing the relationship between the signal and magnetization in the object. We will assume for simplicity that $T_s \ll T_2$ and that the effect of magnetic field inhomogeneities during data acquisition can be neglected. We will also assume that the transmitter and receiver coils used for excitation and signal detection, respectively, generate uniform r.f. fields. From the principle of reciprocity (see Chapter 1), it then follows that the sampled signal in a one-dimensional case can be written as

$$S(n) \equiv S(t(n)) = \xi \int M_{xy}(x) e^{j\phi(x,t(n))} \, dx = \xi \int M_{xy}(x) e^{jk_x(n)x} \, dx. \quad (3.2.5)$$

In this equation ξ is a constant that, for simplicity, will be given the value of one in the derivations below, M_{xy} is the (complex) transverse magnetization in the object (see Appendix), and

$$k_x(n) = \gamma G_{0,x} \tau_w n. \quad (3.2.6)$$

From Eq. (3.2.5) it follows that the acquired signal is defined by Fourier transform of the magnetization in the object. Therefore, according to the results obtained in the previous section, inverse discrete FT of the signal in Eq. (3.2.5) will reconstruct M_{xy} with spatial resolution L_x/N, where

$$L_x = 2\pi/\Delta k_x = 2\pi/(\gamma G_{0,x}\tau_w). \qquad (3.2.7)$$

Sampling

   To understand better the important relationship between the field-of-view, readout gradient, and dwell time in Eq. (3.2.7), we need to recall that according to the sampling theorem [1] a continuous function $f(t)$ can be reconstructed fully from a series of discrete samples provided that: a) $FT\{f(t)\}$ is zero for all frequencies $|\nu| > \nu_{max}$; and b) sampling interval does not exceed $1/2\nu_{max}$. Conversely, if sampling interval exceeds $1/2\nu_{max}$ then some of the high-frequency components of $f(t)$ cannot be determined. The insufficiently high sampling rate in the latter case, often referred to as *undersampling*, causes aliasing, which generally makes it impossible to reconstruct a continuous function faithfully from its discrete samples.

   The signal bandwidth in MR imaging is defined by the readout gradient and field-of-view

$$2\nu_{max} = \frac{\gamma G_{0,x}L_x}{2\pi}. \qquad (3.2.8)$$

The signal is digitized at a rate of τ_w^{-1} by using a special device called analog-to-digital converter (ADC). By examining Eq. (3.2.7) in terms of the sampling theorem, we realize that this equation ensures that the sampling rate is sufficient (i.e., $\tau_w = 1/2\nu_{max}$) to avoid aliasing when reconstructing a continuous NMR signal from its discrete samples.

Phase Encoding

   Two-dimensional spatial encoding in MRI is normally achieved through the use of an additional gradient, known as the *phase-encoding gradient*, which is perpendicular to the frequency-encoding gradient. In the presence of the phase-encoding gradient, $G_y = dB_z/dy$, applied prior to readout, the transverse magnetization acquires a phase

$$\phi(y) = \gamma y \int_t^{t+t_{ph}} G_y\, dt, \qquad (3.2.9)$$

where t_{ph} is the duration of the gradient. As a result of frequency and phase encodings the NMR signal in a two-dimensional case is given by

2D Fourier transform of the magnetization:

$$S = \int\int M_{xy}(x,y)e^{jk_x(n)x+j\phi(y)}\,dx\,dy$$

$$= \int\int M_{xy}(x,y)e^{jk_x(n)x+jk_yy}\,dx\,dy, \qquad (3.2.10)$$

where $k_y = \gamma \int_t^{t+t_{ph}} G_y\,dt$.

Phase encoding requires repetitive excitations of the transverse magnetization in the object in order to collect signals at different values of k_y. In the initially proposed phase-encoding scheme, changes in k_y were achieved by varying the duration of the phase-encoding gradient while keeping its strength constant [4]. A more common approach is to vary the strength of the phase-encoding gradient in a step-like fashion; that is,

$$G_y = G_{0,y}m, \qquad (3.2.11)$$

while its duration, t_{ph}, is kept constant [5]. In Eq. (3.2.11) $G_{0,y}$ is a constant, and an integer m changes from $-M/2$ to $M/2-1$ (M is the total number of phase-encoding steps).

Using Eq. (3.2.11), the signal obtained with frequency and phase encodings can be written as

$$S(n,m) = \int\int M_{xy}(x,y)e^{jk_x(n)x+jk_y(m)y}\,dx\,dy, \qquad (3.2.12)$$

where $k_x(n) = \gamma G_{0,x}\tau_w n$ and $k_y(m) = \gamma G_{0,y}t_{ph}m$. After all phase encodings are implemented, image reconstruction is performed by computing inverse discrete FT of $S(n,m)$. Using results from the previous section we find that the reconstructed image intensity is given by the convolution of the transverse magnetization in the sample and a two-dimensional *PSF*. The resultant spatial resolution in the x-y plane is $(L_x/N) \times (L_y/M)$, where L_y is the field-of-view in the direction of the phase-encoding gradient:

$$L_y = 2\pi/\Delta k_y = 2\pi/(\gamma G_{0,y}t_{ph}). \qquad (3.2.13)$$

Note that the reconstructed image intensity is a complex quantity which can be written in polar form $Ae^{j\theta}$, where A and θ are the amplitude and phase of the intensity, respectively. It is the amplitude of the intensity that is normally used to display MR images (often referred to as *magnitude images*), while the phase information is neglected. However, in several instances (e.g., imaging of flow, imaging of static magnetic field inhomogeneities etc.) the reconstructed phase of the intensity may also be used. Basic aspects of image display are discussed in the Appendix.

3.3. BASIC GRADIENT AND SPIN-ECHO PULSE SEQUENCES

Gradient-Echo Sequence

A gradient-echo pulse sequence begins with excitation of the transverse magnetization in a slice of material. After the excitation the polarity of the slice-select gradient (shown as G_z in Figure 3.3) is reversed in order to refocus spins in the slice (see Chapter 2). The excited slice is subsequently imaged in the x- and y-directions using frequency and phase encodings. The pulse sequence including spatially selective excitation, phase encoding and readout is repeated M times with different values of the phase-encoding gradient as described in the previous section. If TR denotes the time interval between consecutive excitations (repetition time), then the total scan time, T_{scan}, equals $M \times TR$.

Reconstructed image intensity is defined by the sum of signals produced by spins from the volume

$$\delta V = \delta x \times \delta y \times L_{sl}, \tag{3.3.1}$$

where $\delta x = L_x/N$ and $\delta y = L_y/M$, L_{sl} is the slice thickness. This result follows from the fact that image intensity is given by the convolution of the transverse magnetization in a slice and a two-dimensional point-spread function that has widths δx and δy in the x- and y-directions, respectively. The volume δV is known as an *imaging voxel* and the area $\delta x \times \delta y$ is known as a *pixel*.

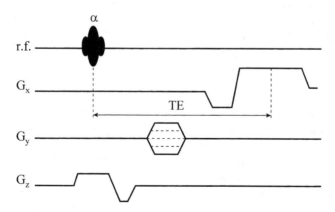

Figure 3.3. Basic gradient-echo sequence. Readout, phase-encoding and slice-select gradients arc shown as G_x, G_y, and G_z, respectively.

Gradient-echo imaging often suffers from signal loss caused by dephasing of spins in the presence of magnetic field inhomogeneities. An empirical formula frequently used to evaluate gradient-echo image intensity is

$$I \propto e^{-TE/T_2^*}, \tag{3.3.2}$$

where *TE*, known as the *echo time*, is the time interval between the center of the excitation r.f. pulse and the gradient echo (see Figure 3.3). Notice that T_2^* is used in Eq. (3.3.2) instead of T_2 in order to take into account the effect of dephasing of spins within a voxel. Since dephasing of spins increases with increasing *TE*, gradient-echo images are often acquired with very short echo time (a few milliseconds) in order to reduce signal loss. The effect of intravoxel dephasing is discussed further in Chapter 6.

Gradient-echo imaging with short *TR* was originally suggested as a technique for rapid *in vivo* MRI in order to decrease scan time and reduce artifacts due to the motion of the object [6–8]. Since its inception rapid gradient-echo imaging has found a number of important clinical applications (e.g., abdominal imaging, MR angiography, and contrast-enhanced MRI). Rapid gradient-echo imaging is discussed further in Chapter 7.

Spin-Echo Sequence

The main difference between gradient-echo and spin-echo pulse sequences is that the latter also includes a 180 degree pulse that causes formation of a spin echo during signal acquisition (Figure 3.4). Initial dephasing of spins in the excited slice can be achieved by applying the defocusing lobe of the readout gradient before the 180 degree pulse. As in gradient-echo imaging, the area of the defocusing lobe equals one-half of the area of the refocusing lobe. The position of the refocusing gradient lobe is typically chosen in such a way that the peak of the spin echo occurs in the center of the acquisition interval. Note that because the 180 degree pulse inverts the phase of the magnetization, the defocusing and refocusing lobes of the readout gradient have the same polarities.

Because of spin refocusing following a 180 degree pulse, spin-echo imaging is generally less sensitive to the presence of magnetic field inhomogeneities than gradient-echo imaging. Spin-echo image intensity can be approximated as follows:

$$I \propto e^{-TE/T_2}, \tag{3.3.3}$$

where *TE* is the time between the excitation pulse and the formed spin echo (Figure 3.4). Equation (3.3.3) shows that, in contrast to

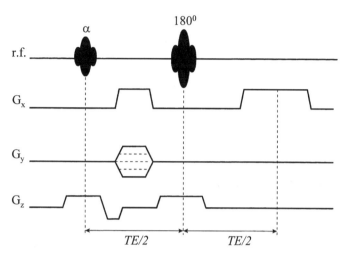

Figure 3.4. Basic spin-echo sequence.

gradient-echo imaging, spin-echo image intensity depends on the ratio of TE to T_2 rather than T_2^*. Because local T_2 values are found to be tissue dependent, spin-echo imaging can provide excellent contrast between different tissues in diagnostic MRI (see Chapter 4).

It should be noted that the use of 180 degree pulses in spin-echo imaging results in longer echo time and repetition time relative to those in gradient echo imaging. As a result spin-echo imaging is generally slower than gradient-echo imaging. An approach that significantly shortens scan time in spin-echo imaging is discussed in Chapter 7.

2D Multi-Slice Imaging

In the case when repetition time TR exceeds the time required for spatially selective excitation, phase encoding, and readout, a number of slices can successively be imaged during one TR interval [9]. Figure 3.5 shows examples of different excitation schemes for multi-slice scanning with two-dimensional frequency and phase encodings, often referred to as conventional or standard MR imaging. The maximum number of imaged slices is given by

$$N_{max} = \frac{TR}{TE + \delta T},\qquad (3.3.4)$$

where δT defines the additional time (approximately half of the readout time) needed to complete signal acquisition and switch gradients. For example, assuming TR of 800 msec, TE of 70 msec, and δT of 10 msec, we obtain $N_{max} = 800/(70 + 10) = 10$ slices.

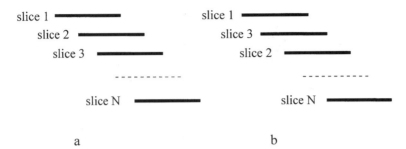

Figure 3.5. Sequential (a) and interleaved (b) excitation schemes for multi-slice imaging.

To prevent artifacts and avoid a reduction in the signal-to-noise ratio in multi-slice imaging, simultaneous excitation of nuclei from different slices should be avoided. This requirement is difficult to satisfy in practice because of imperfections of excitation pulses. Experience shows that signals from neighboring slices often interfere with each other. One solution to the problem of slice interference is to use a gap between imaged slices. Another practical solution is to alternate slices during excitation (Figure 3.5b). The slice interference can also be reduced by using long r.f. pulses, which in general produce better slice profiles than those created by short r.f. pulses.

3D Imaging

The minimum slice thickness in single slice or multi-slice 2D imaging is limited by the maximum gradient strength and duration of the spectrally selective r.f. pulse used for excitation of the transverse magnetization. Typical slice thickness in 2D imaging is about 1–5 mm. Thinner slices are usually acquired by using Fourier encoding in three dimensions (3D FT) [10]. The principal difference between 2D imaging and 3D imaging is that the latter includes an additional phase-encoding gradient that is applied in the direction of the slice-select gradient. A typical 3D FT gradient-echo sequence is shown in Figure 3.6. The spatial resolution in the z direction is L_{sl}/Q, where L_{sl} is the slice thickness and Q is the number of phase-encoding steps in the z-direction. Compared with 2D multi-slice scanning, the 3D approach allows higher spatial resolution and higher signal-to-noise ratio (discussed in Chapter 5). Three-dimensional imaging is normally implemented with relatively thick slabs (i.e., 1–10 cm) while using phase encoding in the slab direction to achieve high spatial resolution. As a result, 3D imaging can be performed with very short excitation

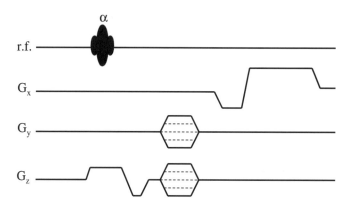

Figure 3.6. Basic gradient-echo sequence for 3D imaging.

pulses, which makes it possible to significantly shorten *TE* and *TR*. This property of 3D imaging is often used in MRI of flow, because short *TE* generally reduces artifacts caused by flow-induced dephasing of spins (see Chapter 6). The major limitation of 3D imaging is its relatively long scan time, T_{scan}, which is defined by the chosen *TR* and the total number of phase encodings: $T_{scan} = M \times Q \times TR$.

3.4. K-SPACE

Tweig [11] and Ljunggren [12] have suggested a very useful interpretation of the signal acquisition in Fourier MRI as sampling along trajectories, $\mathbf{k} = \{k_x(t), k_y(t), k_z(t)\}$, in a spatial frequency space, or *k-space*. The *k*-space trajectories are described by the following equations: $k_x(t) = \gamma \int_0^t G_x \, dt$, $k_y(t) = \gamma \int_0^t G_y \, dt$ and $k_z(t) = \gamma \int_0^t G_z \, dt$. For example, in 2D imaging one of the components of $\mathbf{k}$ produced by the phase-encoding gradient is constant during acquisition, whereas another component produced by the readout gradient changes linearly with time. The corresponding trajectories in *k*-space are represented by a number of parallel lines (Figure 3.7). Several frequently used sampling schemes that allow rapid acquisition of MR images using a small number of *k*-space trajectories such as echo-planar, fast spin-echo and spiral imaging are discussed in Chapter 7.

In general, arbitrary trajectories in *k*-space can be produced by time-varying gradients applied during signal acquisition, assuming compliance with hardware limitations such as maximum available gradient amplitude, G_{max}, and maximum slew rate, $(dG/dt)_{max}$. However, the time during which the signal can be acquired along a

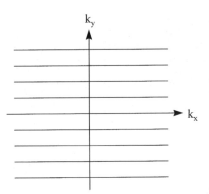

Figure 3.7. K-space trajectories in conventional 2D imaging with frequency and phase-encoding gradients in the x and y directions, respectively.

chosen trajectory is limited by the effects of T_2 decay and magnetic field nonuniformity. In actual practice, these two effects can be significant enough during MR scanning to cause image distortions that can vary from mild to severe depending on the chosen sampling scheme.

The signal amplitude at low spatial frequencies in the central region of k-space is typically much higher than the signal amplitude at high spatial frequencies on the periphery of k-space. This is demonstrated in Figure 3.8, which shows an image of a uniform phantom and the corresponding absolute values of the acquired k-space data. In general, low spatial frequencies are needed to reconstruct the bulk of the object, and high spatial frequencies are required to accurately reproduce rapid spatial variations in the object (e.g., sharp edges). The range of spatial frequencies in k-space is inversely proportional to spatial resolution. For example, in 2D imaging with in-plane resolution $\delta x \times \delta y$ the signal is sampled inside a $(2\pi/\delta x) \times (2\pi/\delta y)$ rectangle in k-space. Because high spatial frequencies beyond a certain limit are not acquired, reconstructed images often suffer from *truncation artifacts* (also known as *Gibbs ringing*) characterized by a series of ripples originating at intensity discontinuities (Figure 3.9).

The appearance of truncation artifacts as ripples in images can be explained by the oscillations of *PSF* (Figure 3.1). In the regions with slow varying magnetization, the result of the convolution of the transverse magnetization in the object and *PSF* will approximate M_{xy} well. In contrast, when sharp spatial changes (discontinuities) in M_{xy} are present, the oscillations in *PSF* will cause oscillations (ripples) in image intensity propagating away from the discontinuities [3]. As

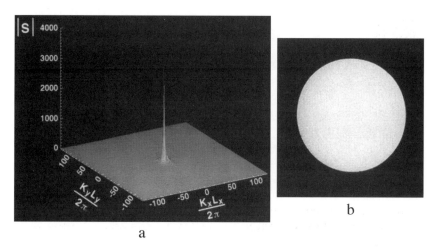

a

Figure 3.8. (a) Plot of the magnitude signal, $|S(k_x, k_y)|$, produced by a homogeneous spherical phantom; (b) reconstructed image of a transverse section of the phantom.

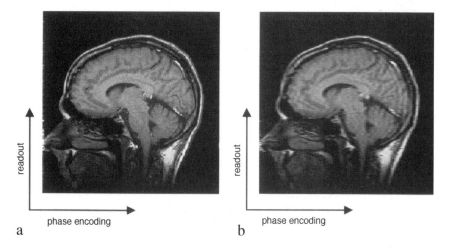

a   b

Figure 3.9. Sagittal images of the head: (a) 256 (readout) × 192 (phase) matrix size and 0.9 mm × 1.25 mm (in-plain) spatial resolution; (b) 256 (readout) × 64 (phase) matrix size and 0.9 mm × 3.75 spatial resolution. The image in (b) displays significant truncation artifacts due to low resolution in the phase direction.

spatial resolution increases ripples with significant amplitudes move closer and closer to discontinuities creating an impression of decreased Gibbs ringing.

REFERENCES

[1] R.N. Bracewell. *The Fourier transform and its applications.* McGraw-Hill (1986).

[2] C.E. Metz, K. Doi. "Transfer function analysis of radiographic imaging systems," *Phys. Med. Biol.* **24**, 1079 (1979).

[3] Z.-P. Liang, F.E. Boada, R.T. Constable, E.M. Haacke, P.C. Lauterbur, M.R. Smith. "Constrained reconstruction methods in MR imaging," *Rev. Magn. Reson. Med.* **4**, 67 (1992).

[4] A. Kumar, D. Welti, R.R. Ernst. "NMR Fourier Zeugmatography," *J. Magn. Reson.* **18**, 69 (1975).

[5] W.A. Edelstein, J.M.S. Hutchison, G. Johnson, T. Redpath. "Spin warp NMR imaging and applications to human whole-body imaging," *Phys. Med. Biol.* **25**, 751 (1980).

[6] A. Haase, J. Frahm, D. Matthaei, W. Hanicke, K.D. Merboldt. "Flash imaging: rapid NMR imaging using low flip angle pulses," *J. Magn. Reson.* **67**, 258 (1986).

[7] P. van Meulen, J.P. Groen, J.J.M. Cuppen. "Very fast MR imaging by field echoes and small angle excitation," *Magn. Reson. Imag.* **3**, 297 (1985).

[8] J.A. Utz, R.J. Herfkens, G. Glover, N. Pelc. "Three second clinical NMR imaging using a gradient recalled acquisition in a steady state mode (GRASS)," *Magn. Reson. Imag.* **4**, 106 (1986).

[9] L. Crooks, M. Arakawa, J. Hoenninger, J. Watts, R. McRee, L. Kaufman, P.L. Davis, A.R. Margulis, J. DeGroot. "Nuclear magnetic resonance whole-body imager operating at 3.5 kGauss," *Radiology* **143**, 169 (1982).

[10] J. Frahm, A. Haase, D. Matthaei. "Rapid three-dimensional MR imaging using the FLASH technique," *J. Comp. Assist. Tomogr.* **10**, 363 (1986).

[11] D.B. Tweig. "The k-trajectory formulation of the NMR imaging process with applications in analysis and synthesis of imaging methods," *Med. Phys.* **10**, 610 (1983).

[12] S. Ljunggren. "A simple graphical representation of Fourier-based imaging methods," *J. Magn. Res.* **54**, 338 (1983).

CHAPTER 4

Contrast in MR Imaging

The observer's ability to differentiate between different structures in images depends on image contrast. MRI has become an important diagnostic tool in medical imaging because it provides the necessary contrast between various soft tissues required to identify pathologic processes. Image contrast, $C(A, B)$, between two different structures, A and B, within the object may be defined as follows:

$$C(A, B) = \frac{|I(A) - I(B)|}{I_{ref}},\qquad (4.0.1)$$

where $I(A)$ and $I(B)$ are the image intensities of A and B, respectively; I_{ref} is an arbitrary reference intensity [1]. In the case of MRI, there is no standard I_{ref}; therefore, the term "image contrast" is often used to characterize the difference between the intensities $I(A)$ and $I(B)$.

Variations in local relaxation times T_1 and T_2, and the amount of hydrogen nuclei per unit volume (known as *proton density* or *N(H)*) are the predominant sources of contrast in proton MRI. The use of local relaxation times for identification of pathology by MRI was inspired by the observation that malignant tissue can have longer T_1 and T_2 than normal tissue [2]. Although the relaxation properties of ^{1}H nuclei in tissue are not completely understood, there is evidence that the pathological states of different tissues are characterized by specific biochemical processes that can alter relaxation times and proton density [3,4].

In this chapter we consider different mechanisms of contrast in ^{1}H MRI as well as techniques used to improve contrast in MR images. The emphasis of the following discussion is on the relationship between image contrast and NMR relevant tissue parameters, such as inherent relaxation times and proton density.

4.1. MAIN CONTRAST MECHANISMS IN ^{1}H MRI

The dependence of MRI contrast on intrinsic relaxation times and proton density can be studied by using the equation describing steady-state transverse magnetization (M_{tr}) in a given pulse sequence. For simplicity, we will focus our discussion on spin-echo imaging with 90 degree excitation pulses. By using Eq. (2.1.6) we obtain

$$M_{tr} = M_0 e^{-TE/T_2}(1 + e^{-TR/T_1} - 2e^{-(TR-TE/2)/T_1}). \qquad (4.1.1)$$

Taking into account that $M_0 \propto N(H)$ and the fact that spin-echo images are typically acquired with $TE \ll 2T_1$, we can express image intensity, I, as

$$I \propto N(H)e^{-TE/T_2}(1 - e^{-TR/T_1}). \qquad (4.1.2)$$

According to this equation, image intensity is directly proportional to proton density, $N(H)$, and T_1 and T_2 related modulations of the intensity are defined by the ratio of TR to T_1 and TE to T_2, respectively. The proton density variations among soft tissues are typically within a few percent (except for brain tissues), whereas T_1 and T_2 relaxation times in different tissues vary by more than 100% (see Table 4.1). It is this large degree of variation in relaxation times that creates the greatest contrast in MRI.

T_1 Contrast

When tissues with different values of T_1 relaxation time experience a train of excitation pulses, tissues with short T_1 will restore their longitudinal magnetization more rapidly following an excitation

Table 4.1. Approximate values of relaxation times at 1.5 Tesla [4].

Tissue	T_1 (ms)	T_2 (ms)
Gray matter	920	101
White matter	790	92
Cerebrospinal fluid (CSF)*	2650	280
Kidney	650	58
Liver	490	43
Skeletal muscle	870	47

* T_1 and T_2 values for CSF are from [1].

pulse and will have greater steady-state transverse magnetization than tissues with long T_1, provided that variations in T_2 and $N(H)$ are sufficiently small. As a result, tissues with short T_1 seem brighter in MR images with dominant T_1 contrast, known as T_1-weighted images, than tissues with long T_1. For example, the relatively low image intensities of the CSF and gray matter in Figure 4.1 are due to their longer T_1 as compared to T_1 in the white matter.

T_1-weighted images are normally acquired with TE less than T_2 in the tissues of interest in order to reduce T_2 modulation of the signal. T_1 contrast can be further enhanced by adjusting the repetition time TR. For example, by examining results from Figure 4.2 we can conclude that T_1 contrast between tissues with close T_1 values diminishes at either very short $TR \ll T_1$ or very long $TR \gg T_1$, whereas the maximum contrast is achieved at the intermediate values of $TR \approx T_1$.

T_2 Contrast

Variations in T_2 relaxation time among different tissues can be used to produce images with predominant T_2 contrast. These images are characterized by the bright appearance of tissues with long T_2 and dark appearance of tissues with short T_2. Because T_2 related modulation of image intensity is defined by the factor $\exp(-TE/T_2)$, it is clear that echo time TE is the adjustable parameter that controls T_2 contrast. In practice, images with predominant T_2 contrast, known as T_2-weighted images, are acquired with TE approximately equal to or longer than the shortest T_2 in the tissues of interest, and with long repetition time (i.e., $TR > T_1$) needed to reduce T_1 modulation of the signal. Figure 4.3 shows T_2-weighted spin-echo images of the brain. Notice that due to the relatively short T_2 of the white matter the transverse magnetization decays more rapidly in the white matter than in the gray matter and CSF. This causes the dark appearance of the white matter in the T_2-weighted images of the brain. On the other hand, the very bright appearance of the CSF in Figure 4.3 is due to its long T_2 and high proton density.

Proton Density Images

Variations in proton density in different tissues can provide additional contrast in MR images. Proton density "weighted" images are produced using short echo time $TE < T_2$ and long repetition time $TR > T_1$ in order to reduce T_2 and T_1 modulations of the signal, respectively. Figure 4.4 demonstrates proton density contrast between the white matter and gray matter in spin-echo imaging of the brain. The apparent

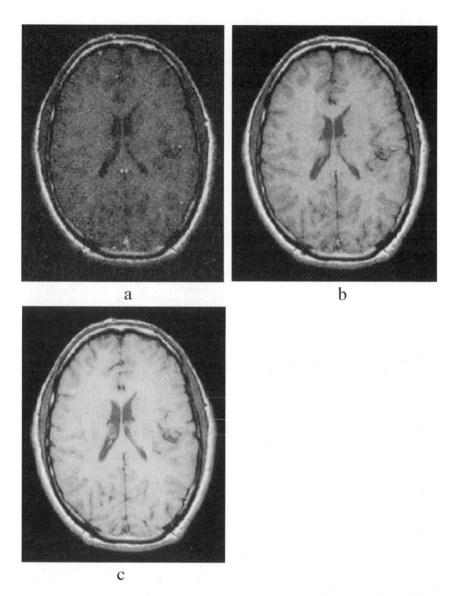

Figure 4.1. T_1-weighted images of the brain: FOV $= 24 \times 24\,\mathrm{cm}^2$, slice thickness $= 2\,\mathrm{mm}$, matrix size $= 256$ (readout) $\times 192$ (phase encoding), $TE = 20\,\mathrm{msec}$. (a) $TR = 150\,\mathrm{msec}$, (b) $TR = 450\,\mathrm{msec}$, (c) $TR = 750\,\mathrm{msec}$. Notice decreased contrast between brain tissues and low signal-to-noise ratio in (a) caused by the increased saturation of the magnetization in these tissues.

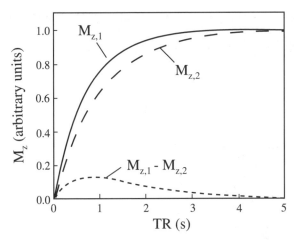

Figure 4.2. Steady-state longitudinal magnetization in spin-echo imaging as a function of TR. $M_{z,1}$ and $M_{z,2}$ correspond to an arbitrary chosen T_1 of 0.7 s and 1 s, respectively.

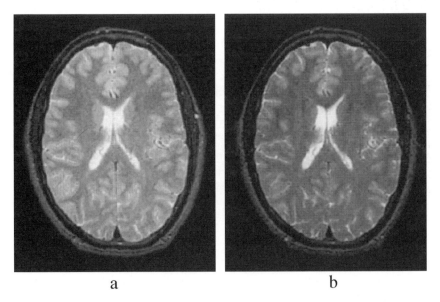

Figure 4.3. T_2-weighted brain images: FOV $= 24 \times 24\,\text{cm}^2$, slice thickness $= 2\,\text{mm}$, matrix size $= 256$ (readout) $\times\ 192$ (phase encoding), $TR = 3\,\text{s}$. (a) $TE = 75\,\text{msec}$; (b) $TE = 150\,\text{msec}$.

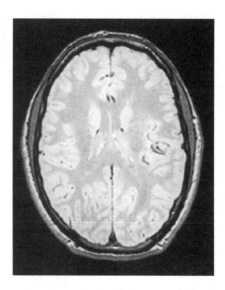

Figure 4.4. Proton density-weighted image of the brain: FOV = 24 × 24 cm^2, slice thickness = 2 mm, matrix size = 256 (readout) × 192 (phase encoding), $TR = 3$ s, $TE = 20$ msec.

low image intensity in white matter is primarily due to its low proton density as compared to gray matter (see Table 4.2). Cerebrospinal fluid, which has higher proton density than that of white matter and gray matter, also has a dark appearance in the image. The explanation of the low image intensity of CSF is that it has a very long T_1 that causes saturation of the CSF even at the relatively long TR of 3 seconds used to acquire the image in Figure 4.4. This example demonstrates that pure proton density contrast among all imaged tissues might be difficult to achieve in practice unless very long TR and short TE are used.

A number of abnormalities such as tumor and edema are characterized by increased values of T_1, T_2, and $N(H)$ relative to their values in

Table 4.2. Relative proton densities in the brain [1].

Tissue	N(H)
Gray matter	0.69
White matter	0.61
Cerebrospinal fluid	1.00
edema	0.86

normal tissue [3]. Spin-echo image intensity decreases with increasing T_1 because of the increased saturation of spins. Conversely, increased T_2 and proton density lead to increased image intensity. Consequently, the appearance of abnormalities in MR images depends on the image weighting. For example, brain tumors can have a dark appearance in T_1-weighted images and a bright appearance in T_2 and proton density–weighted images as compared to surrounding tissue. In practice both T_1-, T_2-, and proton density–weighted images are often acquired in diagnostic MRI in order to improve detection of pathology.

T_2^* Contrast

T_2^*-weighted images are obtained by using gradient-echo imaging with echo time TE comparable to T_2^* in tissue (typically 30–50 ms). T_2^*-weighted images frequently exhibit loss of signal caused by dephasing of spins in a non-uniform magnetic field, and thereby have inferior contrast between soft tissues relative to T_2-weighted spin-echo images. On the other hand, the sensitivity of T_2^*-weighted images to the presence of magnetic field inhomogeneities has proven to be extremely useful for functional MR imaging (FMRI) of the brain.

One of the main physical effects exploited in functional magnetic resonance imaging is the dependence of the magnetic properties of hemoglobin in red blood cells on the blood oxygen level: deoxyhemoglobin is paramagnetic, and oxyhemoglobin is diamagnetic. Ogava and his collaborators [5,6] have demonstrated that the difference in susceptibility between venous blood having a high concentration of deoxyhemoglobin and surrounding brain tissue causes dephasing of nuclear spins, which in turn gives rise to the dark appearance of veins in gradient-echo images. The resulting image contrast is known as *blood oxygen level dependent* contrast (BOLD). FMRI based on BOLD contrast [7–9] employs T_2^*-weighted images to record changes in the signal caused by neuron activation due to various stimuli (e.g., photic stimulation, finger movements etc.). During neuron activation, blood flow in the activated brain regions increases to satisfy the increased demand for oxygen. The hypothesis is that the increase in blood flow is accompanied by an increase in oxygen supply that exceeds the increase in oxygen consumption. Because of the temporary imbalance between the supply and consumption of oxygen, the concentration of deoxyhemoglobin in the activated regions decreases relative to that in the resting state. As a result, the susceptibility difference between blood and its surroundings also decreases, which in turn causes the observed increases in intensity in T_2^*-weighted images of the brain. Since changes in image intensity during activation are small (a few percent at 1.5 Tesla), it is normally necessary to acquire a number of images during the stimulation and resting states and to use statistical analysis to differentiate between the activated and non-activated brain regions.

4.2. CONTRAST AGENTS

Certain materials, known as *contrast agents*, can enhance MR image contrast by altering T_1, T_2, and T_2^* relaxation times. Contrast agents are frequently used in diagnostic MRI in order to achieve better assessment of local physiologic and anatomic conditions or to improve detection of malignancy. Because contrast agents are usually administered internally, they must possess low toxicity and be easily excreted from the body.

Unlike contrast agents used in nuclear medicine and clinical radiography, MRI contrast agents affect the signal indirectly via interaction with hydrogen nuclei. Most of the commonly used MRI contrast agents significantly alter T_1 and T_2 relaxation times in tissue because of dipole-dipole interaction with water protons. Based on the theoretical model by Bloembergen, Purcell, and Pound [10], changes in T_1 and T_2 caused by a contrast agent can be described by the following equations:

$$1/T_1 = 1/T_{1,0} + R_1 n_a,$$
$$1/T_2 = 1/T_{2,0} + R_2 n_a,$$

$$(4.2.1)$$

where T_1 and T_2 are the observed relaxation times; $T_{1,0}$ and $T_{2,0}$ are the relaxation times in the absence of the agent; n_a is the agent's concentration; R_1 and R_2 are the agent's relaxivities.

Paramagnetic ions with a number of unpaired electrons such as Gd^{3+} and Mn^{2+} are frequently used as MRI contrast agents because of their high relaxivities. To reduce the toxicity of these ions for *in vivo* applications, they are usually chelated to special molecules. Gd-DTPA, the best known MRI contrast agent, is a chelate composed of gadolinium ions and diethylenetriaminepentaacetic acid (DTPA). Because chelation of gadolinium ions decreases the effective amount of water molecules that can directly interact with the paramagnetic ions, the R_1 and R_2 relaxivities of Gd-DTPA are smaller than the corresponding relaxivities of free gadolinium ion. The reported R_1 and R_2 relaxivities of Gd-DTPA at 0.5–1.5 Tesla are approximately $4.5 \, kg \cdot mmol^{-1} \cdot s^{-1}$ and $6.0 \, kg \cdot mmol^{-1} \cdot s^{-1}$, respectively [11]. Gd-DTPA is typically administered at a dose of 0.1–0.2 mmol/kg.

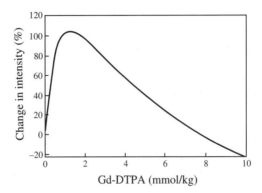

Figure 4.5. Relative change in spin-echo image intensity versus Gd-DTPA concentration. In this example precontrast T_1 and T_2 are 800 msec and 80 msec, respectively. $TR = 400$ msec and $TE = 20$ msec.

Changes in MR image intensity after administration of gadolinium chelates occur due to the effects of T_1 and T_2 shortening [11–13]. These two effects are competing: T_1 shortening leads to increased image intensity and T_2 shortening causes a decrease in intensity. The resulting image intensity is a nonlinear function of contrast agent concentration in tissue and is dependent on other parameters such as intrinsic tissue relaxation times, TR, TE, and flip angle. From the results shown in Figure 4.5, it follows that image intensity initially increases with increasing concentration of Gd-DTPA because of the dominant effect of T_1 shortening. As the Gd-DTPA concentration continues to increase, the effect of T_2 shortening eventually becomes dominant causing a net reduction in image intensity.

Gadolinium-enhanced MRI has been used successfully for the detection of tissue abnormalities. For example, in the brain, administered Gd-DTPA usually remains in the vasculature in normal tissue but can extravasate into the interstitial space in malignant tumors because of the locally disrupted blood-brain barrier [14]. This allows detection of brain tumors based on the observed intensity changes in post-contrast brain images. Another example is contrast-enhanced MRI of the breast. A number of studies have demonstrated intense, rapid signal enhancement in malignant breast tumors as compared to the enhancement in normal tissue [15,16]. Figure 4.6 shows pre- and post-contrast gradient-echo images of a patient with breast carcinoma. Notice significant signal enhancement in the tumor after administration of Gd-DTPA, whereas only relatively small changes are noticeable in

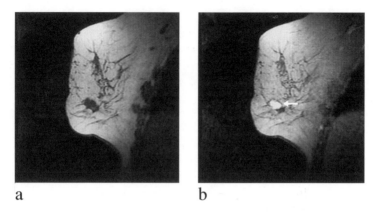

a   b

Figure 4.6. Contrast-enhanced MRI of the breast. The patient is a 74-year-old woman diagnosed with infiltrating ductal carcinoma of the right breast: (a) pre-contrast and (b) post-contrast sagittal MR images of the right breast. Imaging parameters: $FOV = 18 \times 18\,\mathrm{cm}^2$, slice thickness = 2 mm, imaging matrix = 256×128, $TR = 20\,\mathrm{msec}$, $TE = 4\,\mathrm{msec}$, flip angle = 45 degrees. Tumor lesion in (b) is indicated by an arrow.

normal tissue. Other clinical applications of gadolinium-based contrast agents include cardiac, vascular, and perfusion MR imaging [14].

Susceptibility Agents

Contrast agents that significantly alter the local magnetic field in the specimen are frequently referred to as *susceptibility agents*. Such agents typically include different iron oxides [17]. Macroscopic samples of iron are characterized by the existence of magnetic domains. In their natural state the magnetic moments of these domains are oriented randomly, but in the presence of an external magnetic field, magnetic moments are aligned along the field. Iron oxide crystals used as MR contrast agents are extremely small (<50 nm); because of their small size, individual crystals usually contain a single magnetic domain. Similar to ferromagnetic materials, single-domain particles align their magnetic moments along an external field. As a result, magnetic susceptibility in a system of such particles usually far exceeds the susceptibility of paramagnetic materials. On the other hand, similar to paramagnetic materials, the net magnetization in a system of single-domain particles vanishes after the magnetizing field is removed. To characterize the unique magnetic properties of the single-domain crystals of iron oxides, they are often referred to as *superparamagnetic*.

Susceptibility agents are characterized by relatively small R_1 relaxivity compared to paramagnetic chelates. The main effect produced by susceptibility agents is markedly increased local magnetic field gradients, resulting in significantly reduced T_2^* relaxation time and increased attenuation of the NMR signal due to diffusion [18,19].

4.3. MAGNETIZATION TRANSFER CONTRAST

In 1989, Wolff and Balaban discovered a novel form of MRI contrast known as *magnetization transfer contrast* (or MTC). MTC is based on the effect of magnetization transfer between two distinct pools of protons found in biological tissues [20,21]. These pools can be described as: a) freely mobile water protons (^1H_f) characterized by relatively long T_2 (>10 ms); and b) restricted motion macromolecular protons (^1H_r) in proteins and lipids with very short T_2 (<100 μs) (Figure 4.7). The magnetization transfer effect that exists due to dipole-dipole interaction and chemical exchange between the mobile and macromolecular protons, can be described by the modified Bloch equations for the longitudinal magnetizations of the ^1H_f and ^1H_r pools [22,23]:

$$\frac{dM_{z,f}}{dt} = \frac{M_{0,f} - M_{z,f}}{T_{1,f}} - k_f M_{z,f} + k_r M_{z,r},$$

$$\frac{dM_{z,r}}{dt} = \frac{M_{0,r} - M_{z,r}}{T_{1,r}} - k_r M_{z,r} + k_f M_{z,f},$$

$$(4.3.1)$$

where indices f and r denote mobile and restricted motion protons, respectively; M_0 is the equilibrium magnetization; k_f and k_r are the relaxation rate constants describing the coupling between the two pools of protons.

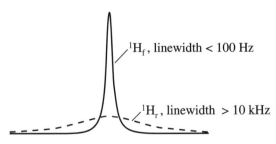

^1H_f, linewidth < 100 Hz

^1H_r, linewidth > 10 kHz

Figure 4.7. Schematic of ^1H_f and ^1H_r tissue spectra.

The main contribution to the observed NMR signal comes from the mobile protons; the signal from the macromolecular protons is normally negligibly small due to their short T_2. Magnetization transfer affects the signal indirectly via interaction between the two pools of protons. In general, attenuation of the signal depends on the degree of saturation of the ^1H_r pool. However, the basic features of MTC can be described by assuming completely saturated macromolecular protons (i.e., $M_{z,r} = 0$). In the latter case we obtain from Eq. (4.3.1)

$$\frac{dM_{z,f}}{dt} = \frac{M_{0,f} - M_{z,f}}{T_{1,f}} - k_f M_{z,f}. \tag{4.3.2}$$

The solution of this equation is given by

$$M_{z,f} = M_{0,f} \frac{T_{1,app}}{T_{1,f}} (1 - e^{-t/T_{1,app}}) + M_{z,f}(t = 0)e^{-t/T_{1,app}}, \tag{4.3.3}$$

where $T_{1,app} = T_{1,f}/(1 + k_f T_{1,f})$. Equation (4.3.3) shows that longitudinal relaxation of $M_{z,f}$ in the presence of magnetization transfer is defined by the apparent relaxation time $T_{1,app}$, which is shorter than $T_{1,f}$. The exchange between the two pools also results in a decreased value of the asymptotic magnetization:

$$M_{s,f} = M_{0,f} \frac{T_{1,app}}{T_{1,f}} = \frac{M_{0,f}}{1 + k_f T_{1,f}}, \tag{4.3.4}$$

where $M_{s,f} = M_{z,f}(t \to \infty)$.

Using equation (4.3.3) in the case of repetitive r.f. excitations with flip angle α and repetition time TR (considered previously in Chapter 2), we can obtain the following equation for the steady-state longitudinal magnetization (see Eq. (2.1.3)):

$$M_{z,f}(TR) = M_{s,f} \frac{(1 - e^{-TR/T_{1,app}})}{(1 - e^{-TR/T_{1,app}} \cos \alpha)}. \tag{4.3.5}$$

From this equation it follows that the signal at time TE after excitation can be expressed as

$$S \propto M_{s,f} e^{-TE/T_{2,f}^*} \sin \alpha \frac{(1 - e^{-TR/T_{1,app}})}{(1 - e^{-TR/T_{1,app}} \cos \alpha)}. \tag{4.3.6}$$

By examining Equations (4.3.4) and (4.3.6), we find that the effect of magnetization transfer on the observed signal depends upon the product of k_f and T_1. This result is illustrated in Figure 4.8, which shows image intensity as a function of flip angle at different values of $k_f T_1$. It is interesting that the Ernst angle that corresponds to the

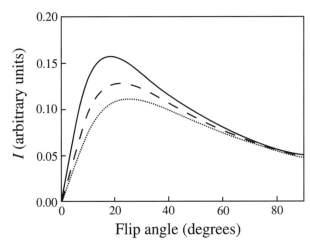

Figure 4.8. Image intensity versus flip angle shown for $T_1/TR = 20$. Solid line: $k_f T_1 = 0$, dashed line: $k_f T_1 = 0.5$, dotted line: $k_f T_1 = 1.0$.

maximum signal from the ^1H_f pool increases in the presence of magnetization transfer due to shortening of the apparent T_1.

Wolff and Balaban, and other investigators have experimentally demonstrated that magnetization transfer between mobile protons and macromolecular protons can significantly decrease NMR signals from certain tissues such as skeletal muscle, and gray and white brain matter [20,24]. Because the magnetization transfer effect is weak in some tissues such as blood, probably due to low concentration of macromolecules, and stronger in others such as muscle, it can enhance contrast in MRI. For example, the magnetization transfer effect can be used to improve visualization of blood vessels in MRI by decreasing the signal from brain parenchyma without significantly affecting the signal from blood [24]. Reference [25] is recommended for a more detailed discussion of MTC and its applications in diagnostic MRI.

MTC in MR imaging is typically achieved by selective saturation of the ^1H_r pool. In the original approach by Wolff and Balaban [20] the ^1H_r pool is saturated by continuous off-resonance r.f. irradiation without affecting the ^1H_f pool. In another technique selective saturation of the ^1H_r pool is achieved by intense on-resonance r.f. pulses [24,26]. A basic 2D gradient-echo pulse sequence utilizing this technique is shown in Figure 4.9. The pulse sequence includes three additional spatially nonselective r.f. pulses that "prepare" magnetization for

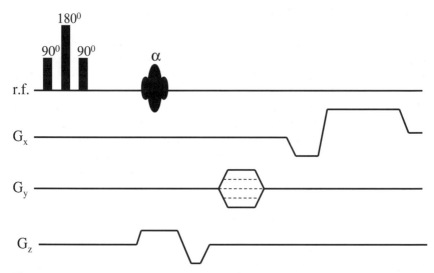

Figure 4.9. A magnetization transfer pulse sequence. The first three pulses providing saturation of the macromolecular protons are followed by a standard 2D gradient-echo pulse sequence.

subsequent imaging. The first 90 degree pulse excites the transverse magnetization in the specimen. The magnetization is subsequently refocused by the 180 degree pulse and eventually rotated back to the z-axis by the second 90 degree pulse. Because of the extremely short T_2 of the macromolecular protons, a short delay ($\sim$1 ms) between the first and second 90 degree pulses is sufficient to greatly decrease the magnetization of the ^1H_r pool without significantly affecting the magnetization of the ^1H_f pool.

Recently Dixon *et al.* demonstrated that MTC can be observed in conventional multislice imaging [27]. To explain the observed effect we need to remember that spatially-selective excitation generates the transverse magnetization for the ^1H_f pool only within the imaged slice without affecting the mobile protons in other slices. However, because of the extremely broad ^1H_r resonance the excitation pulse causes saturation of the macromolecular protons in these slices. As a result, the mobile water protons lose some of their longitudinal magnetization via interaction with the macromolecular protons, and subsequent excitations of these slices therefore produce less signal. The resultant effect is a small decrease in intensity in multislice imaging as compared to single slice and 3D imaging [27].

4.4. DIFFUSION-WEIGHTED IMAGES

An additional attenuation of the signal in proton MRI results from diffusion of water molecules (see Chapter 1). Different microscopic structures such as cell membranes play the role of microbarriers that can significantly alter random motion of water molecules in tissue. Because of these microbarriers diffusion of water molecules in tissue is usually restricted to a limited region in contrast to self-diffusion in pure water. This explains why the reported diffusion coefficients in different tissues are 2 to 3 times smaller than the diffusion coefficient in water ($D \approx 2.2 \times 10^{-5}$ cm^2/s at 25°C). Because of the restricted diffusion the mean square displacement, $\langle r^2 \rangle$, of water molecules in tissue does not increase linearly with time as predicted by the Einstein equation, $\langle r^2 \rangle = 6Dt$. Since the probability of encountering microbarriers increases with time, the diffusion coefficient tends to decrease at longer times. Moreover, different tissue microstructures often exhibit anisotropy, which in turn results in anisotropy of diffusion (i.e., dependence of diffusion on direction) [28,29].

Since signal attenuation due to diffusion is tissue-dependent, it can provide useful image contrast in MRI. Diffusion-weighted imaging (DWI) is typically performed by using the Stejskal-Tanner method [30]. A basic DWI spin-echo pulse sequence shown in Figure 4.10 incorporates two strong gradients providing diffusion-related attenuation of the

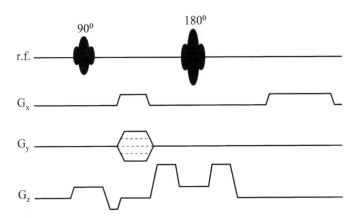

Figure 4.10. 2D DWI spin-echo sequence with diffusion-weighting provided by two strong gradients applied in the z-direction symmetrically with respect to the refocusing 180 degree pulse. In principle, diffusion-sensitizing gradients can be applied in any direction.

signal. Diffusion weighting of the signal in the Stejskal-Tanner sequence is given by the factor e^{-bD}, where b is dependent on the gradient strength, duration and the time interval between the gradients (see Chapter 1). Therefore, when image contrast is primarily defined by diffusion, tissues with large values of D have darker image appearance than tissues with smaller values of D.

MR images acquired with different diffusion weighting can be used for *in vivo* quantitative measurements of the diffusion coefficient in different tissues. Since in practice it is often difficult to discriminate between diffusion and other motions (e.g., blood flow) also contributing to signal attenuation, the diffusion coefficient measured *in vivo* can differ from the actual diffusion coefficient. To distinguish between the measured and the actual diffusion coefficients, the former is often termed *the apparent diffusion coefficient*, or *ADC* [31].

Because it uses strong diffusion gradients, DWI frequently suffers from artifacts caused by eddy currents. The mechanism of these artifacts can be understood from Figure 4.11. Distortions of the gradient waveform caused by eddy currents lead to an additional attenuation of the signal. Since the resulting image distortions vary depending on the strength of the applied gradients, they tend to decrease accuracy in the calculation of the diffusion coefficient from a series of images acquired with different

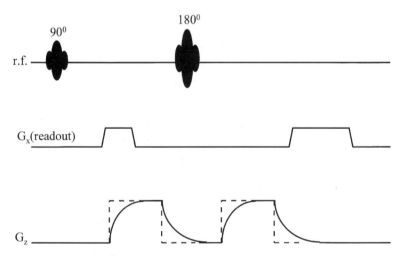

Figure 4.11. Eddy currents produced by switching of strong diffusion-sensitizing gradients cause distortions of the ideal gradient waveform (dashed line). These gradient distortions cause image artifacts and make it more difficult to measure diffusion coefficient accurately.

diffusion gradients. In practice, implementation of DWI normally requires special self-shielded gradient coils or the use of electronic compensation techniques to reduce the effect of eddy currents (see Chapter 9).

In a number of studies it has been suggested that DWI can be used for characterization of tissue perfusion and detection of neurologic disorders [29,31]. Moseley *et al.* [32] have demonstrated that ischemic brain injury is associated with an early and significant reduction in the ADC of water. Because of the changes in ADC, brain regions with a perfusion deficit often seem hyperintense in diffusion-weighted images. This suggests that DWI of the brain may be useful as a diagnostic tool for the early detection of stroke.

REFERENCES

[1] R.E. Hendrick, U. Ruff. "Image contrast and noise, Magnetic Resonance Imaging," D.D. Stark and W.G. Bradly, eds., *Mosby Year Book* (1992).

[2] R. Damadian. "Tumor detection by nuclear magnetic resonance," *Science* **171**, 1151 (1971).

[3] P.A. Bottomley, C.J. Hardy, R.E. Argersinger, G. Allen-Moore. "A review of ^{1}H nuclear magnetic resonance relaxation in pathology: are T_1 and T_2 diagnostic?" *Med. Phys.* **14**, 1 (1987).

[4] P.A. Bottomley, T.H. Foster, R.E. Argersinger, L.M. Pfeifer. "Review of normal tissue hydrogen NMR relaxation times and relaxation mechanisms from 1–100 MHz: dependance on tissue type, NMR frequency, temperature, species, excision and age," *Med. Phys.* **11**, 425 (1984).

[5] S. Ogawa, T.-M. Lee, A.S. Nayak, P. Glynn. "Oxygenation-sensitive contrast in magnetic resonance image of rodent brain at high magnetic fields." *Magn. Reson. Med.* **14**, 68 (1990).

[6] S. Ogawa, T.-M. Lee. "Magnetic resonance imaging of blood vessels at high fields: *in vivo* and *in vitro* measurements and image simulation." *Magn. Reson. Med.* **16**, 9 (1990).

[7] S. Ogawa, T.-M. Lee, A.R. Kay, D.W. Tank. "Brain magnetic resonance imaging with contrast dependent on blood oxygenation." *Proc. Natl. Acad. Sci. USA* **87**, 9868 (1990).

[8] K.K. Kwong, J.W. Belliveau, D.A. Chesler, I.E. Goldeberg, R.M. Weisskoff, B.P. Poncelet, D.N. Kennedy, B.E. Hoppel, M.S. Cohen, R. Turner, H.-M. Cheng, T.J. Brady, B.R. Rosen. "Dynamic magnetic resonance imaging of human brain activity during primary sensory stimulation." *Proc. Natl. Acad. Sci. USA* **89**, 5675 (1992).

[9] P.A. Bendettini, E.C. Wong, R.S. Hinks, R.S. Tikofsky, J.S. Hyde. "Time course EPI of human brain function during task activation." *Magn. Reson. Med.* **25**, 390 (1992).

[10] N. Bloembergen, E.M. Purcell, R.V. Pound. "Relaxation effects in nuclear magnetic resonance absorption," *Phys. Rev.* **73**, 679 (1948).

[11] J.C. Bousquet, S. Saini, D.D. Stark, P.F. Hahn, M. Nigam, J. Wittenberg, J.T. Ferrucci. "Gd-DOTA: characterization of a new paramagnetic complex," *Radiology* **166**, 693 (1988).

[12] G.L. Wolf, E.S. Fobben. "The tissue proton T1 and T2 response to Gadolinium DTPA injection in rabbits. A potential renal contrast agent for NMR imaging," *Investigative Radiology* **19**, 324 (1984).

[13] D.H. Carr, J. Brown, G.M. Bydder, R.E. Steiner, H.J. Weinmann, U. Speck, A.S. Hall, I.R. Young. "Gadolinium-DTPA as a contrast agent in MRI: initial clinical experience in 20 patients," *AJR* **143**, 215 (1984).

[14] A.D. Watson, S.M. Rocklage, M.J. Carvin. "Contrast agents, Magnetic Resonance Imaging," D.D. Stark and W.G. Bradly, eds. *Mosby Year Book* (1992).

[15] W.A. Kaiser, E. Zeitler. "MR imaging of the breast: fast imaging sequences with and without Gd-DTPA," *Radiology* **170**, 681 (1989).

[16] I.S. Gribbestad, G. Nilsen, H.E. Fjosne, S. Kvinnsland, O.A. Haugen, P.A. Rinck. "Comparative signal intensity measurements in dynamic gadolinium-enhanced MR mammography," *JMRI* **4**, 477 (1994).

[17] A.K. Fahlvik, J. Klaveness, D.D. Stark. "Iron oxides as MR imaging contrast agents," *JMRI* **3**, 187 (1993).

[18] P.A. Hardy, R.M. Henkelman. "Transverse relaxation rate enhancement caused by magnetic particulates," *Magn. Reson. Imaging* **7**, 265 (1989).

[19] R.P. Kennan, J. Zhong, J.C. Gore. "Intravascular susceptibility contrast mechanisms in tissues," *Magn. Reson. Med.* **31**, 9 (1994).

[20] S.D. Wolff, R.S. Balaban. "Magnetization transfer contrast (MTC) and tissue water proton relaxation *in vivo*," *Magn. Reson. Med.* **10**, 135 (1989).

[21] H.T. Edzes, E.T. Samulski. "Cross-relaxation and spin diffusion in the proton NMR of hydrated collagen," *Nature* **265**, 521 (1977).

[22] S. Forsen, R.A. Hoffman. "Study of moderately rapid chemical exchange reactions by means of nuclear magnetic double resonance," *J. Chem. Phys.* **11**, 2892 (1963).

[23] G.B. Pike. "Pulsed magnetization transfer contrast in gradient echo imaging: a two-pool analytic description of signal response," *Magn. Reson. Med.* **36**, 95 (1996).

[24] G.B. Pike, B.S. Hu, G.H. Glover, D. R. Enzmann. "Magnetization transfer time-of-flight magnetic resonance angiography," *Magn. Reson. Med.* **25**, 372 (1992).

[25] R. Balaban, T.L. Ceckler. "Magnetization transfer contrast in magnetic resonance imaging," *Magn. Reson. Quarterly* **8**, 116 (1992).

[26] B.S. Hu, S.M. Conolly, G.A. Wright, D.G. Nishimura, A. Macovski. "Pulsed Saturation Transfer Contrast," *Magn. Reson. Med.* **26**, 231 (1992).

[27] W.T. Dixon, H. Engels, M. Castillo, M. Sardashti. "Incidental magnetization transfer contrast in standard multislice imaging," *Magn. Reson. Imaging* **8**, 417 (1990).

[28] G.G. Cleveland, D.C. Chang, C.F. Hazlewood, H.E. Rorschach. "Nuclear magnetic resonance measurements of skeletal muscle: anisotrophy of the diffusion coefficient of the intracellular water," *Biophys. J.* **16**, 1043 (1976).

[29] M.E. Moseley, Y. Cohen, J. Kucharczyk, J. Mintorovitch, H.S. Asgari, M.F. Wendland, J. Tsuruda, D. Norman. "Diffusion-weighted MR imaging of anisotropic water diffusion in cat central nervous system," *Radiology* **176**, 439 (1990).

[30] E.O. Stejskal and J.E. Tanner. "Spin diffusion measurements: spin echoes in the presence of a time-dependent field gradient," *J. Chem. Phys.* **42**, 288 (1965).

[31] D. Le Bihan, E. Breton, D. Lallemand, P. Grenier, E. Cabanis, M. Laval-Jeantet. "MR imaging of intravoxel incoherent motions: application to diffusion and perfusion in neurologic disorders," *Radiology* **161**, 401 (1986).

[32] M.E. Moseley, J. Kucharczyk and J. Mintorovitch. "Diffusion-weighted MR imaging of acute stroke: correlation with T2-weigthed and magnetic susceptibility-enhanced imaging in cats," *AJNR* **11**, 423 (1990).

CHAPTER 5

Signal-to-Noise Ratio in MRI

The signal used for reconstruction of MR images is always corrupted by noise caused by randomly fluctuating currents in the receiver coil and the imaged object. The quality of MR images can be severely degraded by noise, making it difficult to distinguish between different structures in the object. The three most important parameters defining image quality are spatial resolution, image contrast, and signal-to-noise ratio (SNR). The first two parameters were discussed in Chapters 3 and 4. The third parameter, SNR, is defined as the ratio of the mean image intensity in a chosen region of interest (*ROI*) to the square root of the noise variance:

$$\text{SNR} = \frac{\text{mean intensity}}{\sqrt{\text{noise variance}}}.$$

SNR in MRI depends upon a number of factors including strength of the static magnetic field, type and characteristics of r.f. coils, imaging parameters (e.g., image resolution and matrix size), and chosen pulse sequence. The following discussion focuses on the relationship between SNR and imaging parameters as well as the field strength [1–8]. For a discussion of SNR in different pulse sequences and SNR performance of r.f. coils, see references [1,9–16].

5.1. NOISE VARIANCE IN MR IMAGES

2D Imaging

The induced voltage in the NMR receiver coil is composed mainly of components with frequencies distributed in a narrow range around the frequency $\omega_0 = \gamma B_0$. Acquisition of the NMR signal is normally performed by using phase-sensitive detectors (see Appendix) that shift the signal frequencies by $+/-\omega_0$. The output of a phase-sensitive detector is a sum of low and high frequency components centered at zero and $2\omega_0$, respectively. The high frequency components of the output signal, V, are subsequently removed by passing the signal through a low-pass filter,Φ. This process can be described as a convolution

$$S(t,m) = \int \Phi(t - t')\,V(t',m)dt', \tag{5.1.1}$$

where index m defines a phase-encoding cycle, $m = -M/2, \ldots, M/2 - 1$. By using an analog-to-digital converter the resultant signal, $S(t,m)$, is subsequently sampled at a rate of $1/\tau_w$.

In the presence of noise, the output of a phase-sensitive detector can be written as

$$V(t,m) = V_s(t,m) + \tilde{V}(t,m), \tag{5.1.2}$$

where V_s denotes the true signal and $\tilde{V}(t,m)$ denotes the noise part of the signal. Following phase-sensitive detection, the low-pass filtering of the output signal removes all signal components with frequencies higher than a certain cutoff frequency. To choose the cutoff frequency, we need to determine the range of frequencies in the part of the output signal that consists of low frequency components. In 2D imaging the true signal is described by the following equation (see Chapter 3 and Appendix):

$$V_s(t,m) \propto L_{sl} \iint M_{xy}(x,y)\exp(j\gamma G_x x t + j\gamma G_y(m)y t_{ph})\,dx\,dy. \tag{5.1.3}$$

In this equation M_{xy} is the transverse magnetization; G_x and G_y are the frequency and phase-encoding gradients, respectively; t_{ph} is the duration of the phase-encoding gradient; L_{sl} is the slice thickness.

Under the circumstances typical for 2D imaging the transverse magnetization is zero outside of the chosen field-of-view (FOV), $L_x \times L_y$, in the x–y plane. That is,

$$M_{xy}(x,y) = 0, \text{ if } |x| \geq L_x/2 \quad \text{and/or} \quad |y| \geq L_y/2. \tag{5.1.4}$$

By using Equations (5.1.3) and (5.1.4) it can be shown that the spectral components of the signal V_s are zero for all frequencies $|\nu| > \nu_s/2$, where $\nu_s = 1/\tau_w = |\gamma G_x L_x/2\pi|$ is known as *the sampling bandwidth*. Therefore, the true signal remains unchanged after passing through a low-pass filter that has a cutoff frequency of $\nu_s/2$:

$$\Phi(t) = \frac{\sin \pi \nu_s t}{\pi t}. \tag{5.1.5}$$

In 2D imaging with N readout points and M phase encodings, the noise contribution to a $N \times M$ image intensity can be written as

$$\eta(n, m) = \frac{1}{NM} \sum_{n'} \sum_{m'} e^{-j2\pi nn'/N - j2\pi mm'/M} S_{noise}(n', m'). \qquad (5.1.6)$$

In this equation n and n' change from $-N/2$ to $N/2 - 1$; m and m' change from $-M/2$ to $M/2 - 1$; $S_{noise}(n', m') = \int \Phi(n'\tau_{dw} - t') \tilde{V}(t', m')dt'$ is the noise signal after low-pass filtering and digitization. To calculate the noise variance, we consider white noise that has zero mean and auto-covariance

$$E[\tilde{V}(t', m'), \tilde{V}^*(t'', m'')] = \begin{cases} \sigma^2 \delta(t' - t''), & m' = m'' \\ 0, & m' \neq m'' \end{cases} \qquad (5.1.7)$$

where $E[x(t)]$ is the expected value of $x(t)$, σ is a constant, and δ is the Dirac delta function. From Equations (5.1.5–5.1.7) it follows that the noise variance in an image can be expressed as

$$Var(\eta) \equiv E[|\eta(n, m)|^2]$$

$$= \frac{\sigma^2}{N^2 M} \sum_{n', n''} \exp(-j2\pi n(n' - n'')/N) \frac{\sin \pi \nu_s \tau_w (n' - n'')}{\pi \tau_w (n' - n'')}. \qquad (5.1.8)$$

Taking into account that the interval between sampling points, τ_w, is the inverse of the sampling bandwidth, ν_s, we finally obtain

$$Var(\eta) = \frac{\sigma^2 \nu_s}{N \cdot M} = \frac{\sigma^2}{M \cdot T_s}, \qquad (5.1.9)$$

where $T_s = N/\nu_s$ is the total acquisition time. Therefore, the noise level measured by $\sqrt{Var(\eta)}$ is inversely proportional to the square root of the acquisition time. Notice that the noise variance in Eq. (5.1.9) is spatially independent. In principle this allows measurements of $Var(\eta)$ in an arbitrary region within the field-of-view if this region contains a sufficient number of pixels making statistical averaging possible. In actual practice it is often convenient to measure noise in the background outside of the object's boundaries where noise is the only contributor to image intensity.

3D Imaging

In the case of 3D imaging with N readout points and a total of $M \times Q$ phase encodings, the output of the phase-sensitive detector can be written as

$$V(t, m, q) = V_s(t, m, q) + \tilde{V}(t, m, q). \qquad (5.1.10)$$

In this equation index q indicates an additional phase-encoding used for 3D imaging: $q = -Q/2, \ldots, Q/2 - 1$ (see Chapter 3). After low-pass filtering and digitization, the noise signal is given by

$$S_{noise}(n', m'q') = \int \Phi(n'\tau_{dw} - t')\, \tilde{V}(t', m', q')dt'. \qquad (5.1.11)$$

The noise part of a $N \times M \times Q$ image is given by a three-dimensional Fourier transform of S_{noise}:

$$\eta(n, m, q) = \frac{1}{NMQ}\sum_{n'}\sum_{m'}\sum_{q'} e^{-j2\pi nn'/N - j2\pi mm'/M - j2\pi qq'/Q} S_{noise}(n', m', q').$$

$$(5.1.12)$$

In the case of 3D imaging, the noise autocovariance can be expressed as

$$E[\tilde{V}(t', m', q'),\, \tilde{V}^*(t'', m'', q'')] = \begin{cases} \sigma^2 \delta(t' - t''), & m' = m'', q' = q'' \\ 0, & m' \neq m'' \text{ and/or } q' \neq q'' \end{cases}.$$

$$(5.1.13)$$

By using Equations (5.1.11) through (5.1.13) it can be shown that the noise variance in 3D imaging is given by

$$Var(\eta) = \frac{\sigma^2 \nu_s}{N \cdot M \cdot Q} = \frac{\sigma^2}{M \cdot Q \cdot T_s}. \qquad (5.1.14)$$

5.2. SNR IN 2D AND 3D MRI

Based on the results of the previous section, we consider SNR dependence on the basic imaging parameters such as voxel volume, matrix size, and acquisition time in 2D and 3D imaging. The obtained results demonstrate that 3D imaging in general provides higher SNR as compared to 2D imaging.

2D Imaging

By using the results from Chapter 3 and assuming for simplicity that M_{xy} is uniform within the excited slice we obtain

$$I \propto \frac{M_{xy} \cdot L_{sl} \cdot L_x \cdot L_y}{N \cdot M}, \qquad (5.2.1)$$

where I is the image intensity in the absence of noise and $L_{sl} \cdot L_x \cdot L_y/(N \cdot M)$ is the voxel volume, δV. From Eq. (5.2.1) and

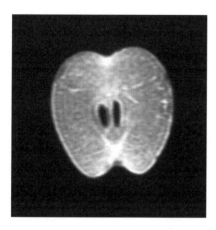 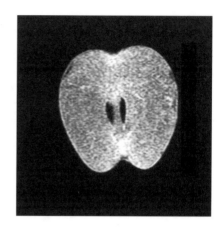

a   b

Figure 5.1. Images of an apple in (a) and (b) were obtained with the same parameters except for FOV, which was reduced by a factor of 4 in (b) as compared to (a). The reduction in FOV decreased the SNR and lowered the overall image quality in (b).

Eq. (5.1.9) we obtain that the SNR in 2D imaging is given by[1]

$$\text{SNR} \propto M_{xy}\delta V \sqrt{M \cdot T_s}. \tag{5.2.2}$$

This equation shows that SNR is proportional to voxel volume. Although decreasing voxel volume generally improves visualization of small structures in images, it causes a proportional decrease in SNR. It should be noted that the decreased SNR can result in unacceptably low image quality, defeating the initial purpose of the decreased voxel size (Figure 5.1).

The SNR dependence on voxel volume, number of phase-encoding steps, and acquisition time should be kept in mind when choosing imaging parameters. For example, if one decides to decrease L_x and L_y by a factor of 2 (without altering matrix size and T_s) in order to increase in-plane spatial resolution, then the SNR will be decreased by a factor of four. Conversely, if the same increase in spatial resolution is achieved by correspondingly increasing matrix size (e.g., using $N' = 2N$ and $M = 2M$) while maintaining FOV and T_s, then

---

[1] Note that the SNR in Eq. (5.2.2) is a complex quantity. In practice SNR is frequently calculated in magnitude MR images. The characteristics of noise in magnitude images are discussed in [17].

the SNR in the image will be reduced only by a factor of $2\sqrt{2}$ at the expense of a twofold increase in scan time.

3D Imaging

Image intensity in 3D MRI is proportional to the product of M_{xy} and the voxel volume, $\delta V = L_x \cdot L_y \cdot L_{sl}/(N \cdot M \cdot Q)$. By using Eq. (5.1.14), the expression for the SNR in 3D imaging can be written as

$$\text{SNR} \propto M_{xy}\delta V\sqrt{M \cdot Q \cdot T_s}. \tag{5.2.3}$$

After comparing Equations (5.2.3) and (5.2.2) one can conclude that the SNR in 3D imaging is $\sqrt{Q}$ times greater than the corresponding SNR in 2D imaging. This result is due to more statistical averaging of the noise in 3D imaging as compared to 2D imaging. The actual increase in SNR in 3D imaging can be very significant. For example, when imaging with the same voxel volume and acquisition time, the SNR in 3D imaging with $Q = 64$ will be eight times greater than the SNR in 2D imaging.

The Effect of Signal Averaging on SNR

Suppose that *NSA* identical (except for noise) frequency and phase-encoded signals are acquired and subsequently averaged together. Because averaging reduces the noise variance by a factor of *NSA* without altering the true signal, it is clear that averaging increases SNR by a factor of $(NSA)^{1/2}$.

The number of times the signal can be averaged in 2D imaging is given by

$$NSA \propto \frac{N'_{sl} \cdot T_{scan}}{N_{sl} \cdot M \cdot TR}, \tag{5.2.4}$$

where T_{scan} is the total scan time, TR is the repetition time, M is the number of phase-encoding steps, N_{sl} is the number of slices, and N'_{sl} is the number of slices imaged during one TR interval. By multiplying the expression for the SNR in Eq. (5.2.2) by $(NSA)^{1/2}$ we obtain

$$\text{SNR} \propto \frac{M_{xy}\delta V\sqrt{N'_{sl} \cdot T_s \cdot T_{scan}}}{\sqrt{N_{sl} \cdot TR}}. \tag{5.2.5}$$

We would like to compare the SNR in 3D imaging with that in 2D imaging under the condition that the scan time is the same. The number of times the signal can be averaged in 3D imaging is

$$NSA \propto \frac{T_{scan}}{M \cdot Q \cdot TR}. \tag{5.2.6}$$

By combining Equations (5.2.6) and (5.2.3), the SNR in 3D imaging can be expressed as

$$\text{SNR} \propto \frac{M_{xy}\delta V \sqrt{T_s \cdot T_{scan}}}{\sqrt{TR}}. \tag{5.2.7}$$

In summary, the SNR in 2D and 3D imaging is proportional to the voxel volume and the square root of the acquisition time. Three-dimensional imaging in general has higher SNR than 2D imaging given the same M_{xy}, voxel size and acquisition time. The SNR in 2D and 3D imaging increases as the square root of *NSA* at the expense of prolonged scan time (which is proportional to *NSA*). After examining Eqs. (5.2.5) and (5.2.7), we conclude that in the presence of signal averaging, the SNR in 2D imaging approaches that in 3D imaging as the number of slices imaged during a single *TR* interval, N'_{sl}, approaches the total number of slices, N_{sl}.

5.3. SNR FIELD DEPENDENCE

Experience shows that SNR in MR images varies depending on the strength of the static magnetic field, B_0. Because of its importance for MRI, the SNR dependence on the field strength has been studied by many investigators [1–4]. The following approach, used to derive the SNR field dependence, mainly employs the arguments put forward by Hoult *et al.* [1,2].

To describe the SNR dependence on B_0, we need to consider how the NMR signal and noise variance depend upon B_0. As shown in Chapter 1, the signal is proportional to the square of B_0. By examining, Equations (5.1.9) and (5.1.14) the reader can realize that the dependence of the noise variance on B_0 is "hidden" in the constant σ.

Because the source of noise in MRI is randomly fluctuating currents in the sample and the receiver coil, σ is given by the Johnson formula:

$$\sigma^2 = 4kT(R_{sample} + R_{coil}). \tag{5.3.1}$$

In this equation, k is Boltzmann's constant, T is the temperature, and R_{sample} and R_{coil} are the sample and coil resistances, respectively. Due to the "skin effect," the coil resistance is proportional to the square root of B_0 [1]. To determine the field dependence of R_{sample} we need to consider thermal losses in the sample.

The average power of energy losses in the sample is given by

$$P = \int_{sample} \chi_s \langle E^2 \rangle dV, \tag{5.3.2}$$

where χ_s is the conductivity of the sample, E is the induced electric field, and $\langle E^2 \rangle$ denotes the average value of E^2. On the other hand, the expression for P can also be written as

$$P = \langle I^2 \rangle R_{sample}, \tag{5.3.3}$$

where I is the oscillating current used for excitation of the transverse magnetization in the sample, and $\langle I^2 \rangle$ denotes the average value of I^2. Assuming that $I = I_0 \cos \omega t$, the induced magnetic field in the sample is

$$B \propto I_0 \cos \omega t. \tag{5.3.4}$$

From the equation $\nabla \times \mathbf{E} = -\dfrac{\partial \mathbf{B}}{c\, \partial t}$ it follows that the induced electric field is

$$E \propto \omega I_0 \sin \omega t. \tag{5.3.5}$$

Therefore, $\langle E^2 \rangle$ is proportional to the square of the r.f. frequency ω. At resonance ω is equal to the Larmor frequency of spins γB_0. According to Equations (5.3.2) and (5.3.5), in this case the average losses in the sample are proportional to the square of B_0; therefore $R_{sample} \propto B_0^2$.

The resultant dependence of the noise variance on B_0 can be described as

$$Var(\eta) \propto \alpha_s B_0^2 + \alpha_c B_0^{1/2}, \tag{5.3.6}$$

where α_s and α_c are two constants that depend on the geometry of the sample and the coil, and their conductivities. Because the true signal is proportional to B_0^2, the SNR dependence on the field strength is given by [1]

$$\mathrm{SNR} \propto \frac{B_0^2}{\sqrt{\alpha_s B_0^2 + \alpha_c B_0^{1/2}}}. \tag{5.3.7}$$

This equation shows that thermal noise at low magnetic field strength is primarily defined by the resistance of the coil and the corresponding $\mathrm{SNR} \propto B_0^{7/4}$. Conversely, as B_0 increases, resistance of the sample eventually exceeds that of the coil[2]. In the latter case the SNR becomes proportional to B_0.

It should be noted that there exist several additional factors which can influence the SNR dependence on B_0. First, because T_1 relaxation time decreases with decreasing field strength [18], SNR at low field

---

[2] The majority of whole body MR scanners operate at field strengths between 0.3 and 2 Tesla. The reported measurements of the SNR field dependence [3] clearly indicate that for these field strengths $R_{sample} > R_{coil}$.

strength benefits from the smaller saturation of imaged material relative to that at high field strength. Second, the dependence of the sample conductivity on r.f. frequency becomes important at high field strengths when SNR $\propto B_0/\sqrt{\chi_s}$. That is, according to [19] changes in the tissue conductivity alter the linear relationship between SNR and field strength at B_0 greater than 1.5 Tesla. Another factor, which becomes important when imaging at high field strength, is that penetration of r.f. field inside the object decreases with increasing B_0. The effect of a decreased r.f. penetration on image intensity and SNR is discussed in [20].

REFERENCES

[1] D.I. Hoult, P.C. Lauterbur. "The sensitivity of the zeugmatographic experiment involving human samples," *J. Magn. Reson.* **34**, 425 (1979).

[2] D.I. Hoult, R.E. Richards. "The signal-to-noise ratio of the nuclear magnetic resonance experiment," *J. Magn. Reson.* **24**, 71 (1976).

[3] W.A. Edelstein, G.H. Glover, C.J. Hardy, R.W. Redington. "The intrinsic signal-to-noise ratio in NMR imaging," *Magn. Reson. Med.* **3**, 604 (1986).

[4] C.N. Chen, V.J. Sank, S.M. Cohen, D.I. Hoult. "The field dependence of NMR imaging: 1. Laboratory assessment of signal-to-noise ratio and power deposition," *Magn. Reson. Med.* **3**, 722 (1986).

[5] F.W. Wehrli. "Signal-to-noise and contrast in MR imaging, NRM in Medicine: The instrumentation and clinical applications," S.R. Thomas, R.L. Dixon, eds. *AAPM* (1986).

[6] A. Zakhor, R. Weisskoff, R. Rzedzian. "Optimal sampling and reconstruction of MRI signals resulting from sinusoidal gradients," *IEEE Trans. Sign. Proc.* **39**, 2056 (1991).

[7] D.L. Parker, G.T. Gullberg. "Signal-to-noise efficiency in magnetic resonance imaging," *Med. Phys.* **17**, 250 (1990).

[8] A. Macovski. "Noise in MRI," *Magn. Reson. Med.* **36**, 494 (1996).

[9] W.A. Edelstein, P.A. Bottomley, H.R. Hart, L.S. Smith. "Signal, noise and contrast in nuclear magnetic resonance imaging," *J. Comput. Assist. Tomogr.* **7**, 391 (1983).

[10] F.W. Wehrli, J.R. MacFall, D. Shutts, R. Breger, R.J. Herfkens. "Mechanisms of contrast in NMR imaging," *J. Comput. Assist. Tomogr.* **8**, 369 (1984).

[11] R.B. Buxton, R.R. Edelman, B.R. Rosen, G.L. Wismer, T.J. Brady. "Contrast in rapid MR imaging: T1- and T2-weighted imaging," *J. Comput. Assist. Tomogr.* **11**, 7 (1987).

[12] R.E. Hendrick, U. Raff. "Magnetic Resonance Imaging: Image contrast and noise," D.D. Stark and W.G. Bradley, eds. *Mosby Year Book* (1992).

[13] N. Pelc. "Optimization of flip angle for T1 dependent contrast in MRI," *Magn. Reson. Med.* **29**, 695 (1993).

[14] C.E. Hayes, W.A. Edelstein, J.F. Schenck. "Radio frequency coils, NMR in medicine: the instrumentation and clinical applications," S.R. Thomas and R.L. Dixon, eds. *AAPM* (1986).

[15] C.E. Hayes, W.A. Edelstein, J.F. Schenck, O.M. Mueller, M. Eash. "An efficient, highly homogeneous radiofrequency coil for whole-body NMR imaging at 1.5 T," *J. Magn. Reson.* **63**, 622 (1985).

[16] C.E. Hayes, P.B. Roemer. "Noise correlations in data simultaneously acquired from multiple surface coil arrays," *Magn. Reson. Med.* **161**, 181 (1990).

[17] R.M. Henkelman. "Measurement of signal intensities in the presence of noise in MR images." *Med. Phys.* **12**, 232 (1985).

[18] P.A. Bottomley, T. H. Foster, R.E. Argersinger, L.M. Pfeifer. "Review of normal tissue hydrogen NMR relaxation times and relaxation mechanisms from 1–100 MHz: dependance on tissue type, NMR frequency, temperature, species, excision and age," *Med. Phys.* **11**, 425 (1984).

[19] T.H. Foster. "Tissue conductivity modifies the magnetic resonance intrinsic signal-to-noise ratio at high frequencies." *Magn. Reson. Med.* **23**, 383 (1992).

[20] G.H. Glover, C.E. Hayes, N.J. Pelc, W.A. Edelstein, O.M. Mueller, H.R. Hart, C.J. Hardy, M. O'Donnel, W.D. Barber. "Comparison of linear and circular polarization for magnetic resonance imaging." J. Magn. Reson. **64**, 255 (1985).

CHAPTER 6

Image Artifacts

Experience shows that MR images are degraded frequently by different artifacts that occur for a variety of reasons. Historically, the quality of MR images was often low because of the artifacts caused by heterogeneity of the main field, eddy currents, nonlinearity of imaging gradients and other scanner-related problems. More recently, the frequency of occurrence and severity of these artifacts in MR images have been reduced significantly due to technical improvements in MRI hardware. A detailed description of hardware-related artifacts is beyond the scope of this book. Instead, the following discussion focuses on several major artifacts that are caused by factors predominantly associated with the imaged object: magnetic field nonuniformity, fat/water chemical shift, flow and motion.

6.1. ARTIFACTS DUE TO MAGNETIC FIELD NONUNIFORMITY

During MR imaging the magnetic field in the object can be nonuniform for various reasons including nonuniformity of the external magnetic field, presence of metal implants and spatial variations in magnetic susceptibility of the object. Magnetic field nonuniformity causes two major types of artifacts in MR images: geometric distortion and signal loss due to intravoxel dephasing [1–7].

Geometric Distortion

To describe geometric distortion caused by magnetic field nonuniformity, it is convenient to express the magnetic field in the specimen as a sum of two terms:

$$\mathbf{B} = \mathbf{B}_0 + \mathbf{B}', \tag{6.1.1}$$

where $\mathbf{B}_0$ is uniform and $\mathbf{B}'$ is a function of coordinates. Because MR imaging is typically performed in a very homogeneous magnetic field such that $|\mathbf{B}'| \ll |\mathbf{B}_0|$, the components of $\mathbf{B}'$ perpendicular to $\mathbf{B}_0$ can be neglected because they cause only a minor perturbation of Larmor frequencies of spins. Thus, we can assume that $\mathbf{B}' = \mathbf{k}B'$, where $\mathbf{k}$ is the unit vector in the direction of $\mathbf{B}_0$.

For simplicity we will confine our discussion to the case of 2D gradient-echo imaging considered in Chapter 3. The magnetic field $\mathbf{B}'$ gives rise to the spatially-varying phase of the transverse magnetization $\phi = \omega' t$, where $\omega' = \gamma B'$. A general equation describing the signal acquired in the presence of T_2 relaxation and dephasing of spins caused by the field inhomogeneity $\mathbf{B}'$ can be written as

$$S(k_x, k_y) = \iint M_{xy}(x,y) e^{jk_x x + jk_y y + j\omega'(t+TE) - (t+TE)/T_2} \, dx \, dy, \tag{6.1.2}$$

where $k_x = \gamma G_x t$ and $k_y = \gamma G_y t_{ph}$; G_x and G_y are the readout and phase-encoding gradients, respectively; $t = n\tau_w$ and n changes from $-N/2$ to $N/2 - 1$; TE is the echo time (see Chapter 3). To simplify our analysis we neglect T_2 relaxation and assume that B' can be approximated as a sum of constant and linear terms: $B' = \alpha + G'_x x + G'_y y + G'_z z$. Using these assumptions we obtain from (6.1.2)

$$S(k_x, k_y) = \iint \frac{M_{xy}\left(\dfrac{x - \delta B'/G_x}{1 + G'_x/G_x}, y\right)}{1 + G'_x/G_x} e^{jk_x x + jk_y y + j\omega' TE} \, dx \, dy, \tag{6.1.3}$$

where $\delta B' = \alpha + G'_y y + G'_z z$.

Geometric distortion is produced by intrinsic field gradients G'_x, G'_y and G'_z that alter the dependence of Larmor frequencies of spins on their actual locations in the specimen. By examining (6.1.3) we can conclude that the image distortion can be described by the following: spatially varying image displacement in the readout direction by $\delta B'/G_x$, magnification in the readout direction and weighting of the image intensity by the factor $(1 + G'_x/G_x)^{-1}$. The strength of the readout gradient is typically high enough to allow only minor geometrical distortion in conventional MR imaging. However in some cases, non-uniformity of the magnetic field can be large enough (e.g., due to the presence of metal objects [7]) to cause significant degradation of images.

Another artifact caused by magnetic field nonuniformity is distortion of the slice profile. In Chapter 2 we have shown that spatially selective excitation in a uniform magnetic field creates a nonzero transverse magnetization in a slice of material with thickness defined by the bandwidth of the excitation pulse and the applied gradient. Assuming that $2\omega_b$ is the excitation bandwidth and G_z is the slice-select gradient, we obtain that the boundaries of the slice are: $z_1 = -\omega_b/(\gamma G_z)$ and $z_2 = \omega_b/(\gamma G_z)$ (see Chapter 2). In the presence of a magnetic field inhomogeneity, $B'(x, y, z)$, the slice boundaries are defined by the following equations:

$$z_1 = \frac{-\omega_b - \gamma B'(x, y, z_1)}{\gamma G_z} \quad \text{and} \quad z_2 = \frac{\omega_b - \gamma B'(x, y, z_2)}{\gamma G_z}. \quad (6.1.4)$$

From Eq. (6.1.4) it follows that magnetic field nonuniformity gives rise to nonplanar slices of varying thickness. The degree of slice distortion depends on magnetic field nonuniformity and slice-select gradient: a large B' or small G_z can result in significant distortions of the slice profile.

It should be noted that no image distortion due to magnetic field inhomogeneities exists in the phase-encoding direction. Consequently, distortion free images can in principle be acquired by using phase encoding in three orthogonal directions. However, this acquisition technique is rarely used because it is prohibitively slow.

Intravoxel Dephasing

Magnetic field inhomogeneities cause dephasing of nuclear spins during data acquisition. Spin dephasing in turn leads to a loss of NMR signal. The resulting effect is a noticeable reduction in image intensity (Figure 6.1). Mathematically, the signal loss is defined by the factor $e^{j\omega' TE}$ averaged over the voxel volume. It is instructive to consider the effect of intravoxel dephasing in the presence of an isotropic gradient, $G'_x = G'_y = G'_z$, in a cubic voxel, $\Delta^3 x$. In this case the reduction in image intensity is given by the factor

$$\left| \frac{1}{\Delta^3 x} \int e^{j\omega'(\mathbf{r})TE} d\mathbf{r} \right| = \left| \frac{\sin^3(\Delta\omega TE)}{(\Delta\omega TE)^3} \right|, \quad (6.1.5)$$

where $\Delta\omega = \gamma G'_x \Delta x/2$. For example, an isotropic gradient of $1\,\text{ppm/cm}$ causes a loss of more than 50% of the signal in gradient-echo imaging at $1.5\,\text{Tesla}$ with TE of $30\,\text{msec}$ and $2\,\text{mm}^3$ voxel size. The good news is that decreasing voxel size and shortening echo time reduces loss of

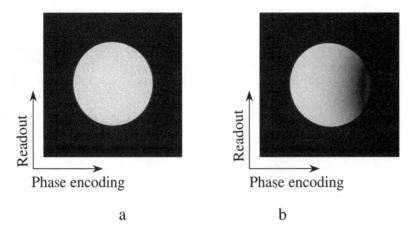

Figure 6.1. Images of a uniform phantom in (a) and (b) were acquired using a gradient-echo pulse sequence with the same imaging parameters. Visible reduction in image intensity in (b) is due to a metal paper clip placed in the vicinity of the phantom.

signal. The bad news is that by decreasing voxel size we proportionally reduce SNR in images.

Equation (6.1.5) is interesting because it clearly indicates that magnetic field inhomogeneities generally cause a nonexponential decay of signal as a function of echo time in gradient-echo imaging. In many instances this decay can be significant enough to cause a drastic reduction in image intensity (Figure 6.1). It is therefore important that the effect of magnetic field nonuniformity can be minimized by acquiring spin-echo signals. Because of the reduced dephasing of spins, the resulting spin-echo images generally have better contrast than gradient-echo images.

6.2. CHEMICAL SHIFT ARTIFACTS

Because of the chemical shift effect, hydrogen nuclei in water and fat have different Larmor frequencies even if the magnetic field in media is highly uniform. Fat/water chemical shift is about 3.5 ppm. For example, at 1.5 Tesla the difference between the resonant frequencies of water, ν_w, and fat, ν_f, is approximately 220 Hz. Chemical shift between fat and water causes two different types of artifacts: (a) misregistration of fat or water in MR images; (b) visible reduction in image intensity as a result of destructive interference between the fat and water signals.

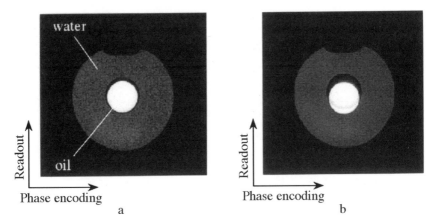

Figure 6.2. Spin-echo images of a phantom composed of a vial with fat-containing vegetable oil placed in a bigger tube filled with water. Images in (a) and (b) were obtained at 1.5 Tesla with readout gradients of 0.6 gauss/cm and 0.15 gauss/cm, respectively. The increased spatial displacement of fat in (b) caused a signal void in the regions that artificially "lost" fat.

As mentioned earlier, by using phase-sensitive detection the frequency of the observed NMR signals is lowered by the reference frequency ω_0. Using Eq. (6.1.3) it can be shown that if the Larmor frequency of water is chosen as the reference frequency, then fat in an MR image will be shifted in the readout direction by

$$\delta x = \frac{2\pi\delta\nu}{\gamma G_x} \qquad (6.2.1)$$

relative to its actual location in the object. In this equation $\delta\nu = \nu_w - \nu_f$ and G_x is the readout gradient. At a field strength of 1.5 Tesla and a readout gradient of 1 gauss/cm, the misplacement of fat is only about 0.5 mm. However, at higher magnetic field strengths or lower gradient strengths the misplacement of fat can become significant enough to cause visible image distortion (see Figure 6.2).

In spin-echo imaging, the acquired phase difference between the water and fat magnetizations is canceled out following a 180° refocusing pulse. As a result water and fat are "in-phase" at echo time TE. In gradient-echo imaging, water and fat can be either in-phase or out-of-phase depending on the product of $\delta\nu$ and TE. For example, the maximum destructive interference between water and fat signals occurs when $TE\delta\nu = (n + 1/2)$, where $n = 0, 1, 2 \ldots$ (Figure 6.3). A common artifact caused by destructive interference between water and fat

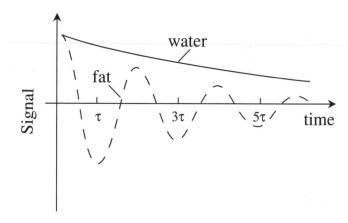

Figure 6.3. As fat and water magnetizations precess with different frequencies, the resulting signal exhibits oscillatory behavior. The signal intensity has its minima at $t = \tau$, 3τ, $5\tau \ldots$, where $\tau = 1/2(\nu_w - \nu_f) \approx 2.2$ msec at 1.5 Tesla.

signals is an apparent signal void in voxels that contain about the same amounts of fat and water. This artifact often gives rise to artificial black boundaries between fat and muscle regions in gradient-echo images (Figure 6.4).

Dixon [8] has suggested a modification of standard spin-echo acquisition in order to obtain images depicting only water or only fat.

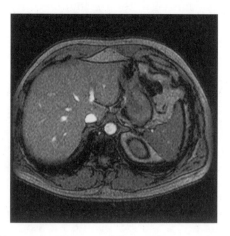

Figure 6.4. Gradient-echo image of the abdomen obtained at 1.5 Tesla with *TE* of 2.2 msec. Boundaries between fat and muscle regions appear black due to interference between the fat and water signals.

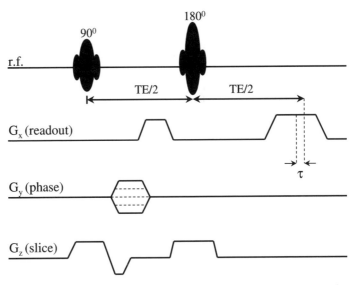

Figure 6.5. Basic pulse sequence in the Dixon method. In one of two images, acquired with $\tau = 0$, water and fat are in-phase. Another image is acquired with $\tau = 1/2\delta\nu$ so that the water and fat magnetizations have opposite directions.

In the Dixon method two images are acquired. One of these images is produced from signals S_{w+f}, obtained with a spin-echo sequence that ensures that signals from water, S_w, and fat, S_f, are in-phase; therefore, $S_{w+f} = S_w + S_f$. Another image is produced from signals, S_{w-f}, acquired during a gradient echo shifted by $\pm 1/2\delta\nu$ relative to the spin echo. In the latter case water and fat are out-of-phase; therefore, $S_{w-f} = S_w - S_f$ (Figure 6.5). Subsequently the water-only image is reconstructed from the sum $S_{w+f} + S_{w-f}$, and the fat-only image is reconstructed from the difference $S_{w+f} - S_{w-f}$ [8,9].

The major problem in the Dixon method is the effect of magnetic field inhomogeneities on the signals from fat and water. In the Dixon method the net signal from water and fat being out-of-phase is given by $S_{w-f} = (S_w - S_f)e^{j\delta\phi}$, where $\delta\phi$ is the phase shift (accumulated during time $1/2\delta\nu$) due to magnetic field inhomogeneities. If $\delta\phi$ is not negligibly small then the Dixon approach can cause artifacts because of incomplete fat/water decomposition.[1]

---

[1] Better differentiation between water and fat can be achieved in the so-called three-point Dixon method [10], which utilizes an additional measurement to separate phase shifts caused by magnetic field nonuniformity from those caused by the fat/water chemical shift.

A technique known as *chemical shift selective* (CHESS) imaging discriminates between fat and water by first selectively exciting and subsequently destroying the magnetization of the unwanted spectral component (e.g., fat) [11]. For example, the fat magnetization can be excited (without affecting the water magnetization) by a frequency-selective 90 degree pulse with narrow bandwidth applied in the absence of external field gradients. After the excited magnetization in fat is destroyed by applying spoiling gradients, the water magnetization is imaged.

Another technique for selective excitation of fat or water employs a family of pulses, known as *binomial pulses.* A binomial pulse can be defined as a combination of equally spaced pulses such as 1–1 or 1–$\bar{1}$, 1–2–1 or 1–$\bar{2}$–1, where the numbers indicate the relative pulse flip angles and the bar over a number indicates a 180° phase-shifted pulse [12,13]. For simplicity we consider selective excitation of fat by a pulse sequence 1–$\bar{1}$, known as the *jump-and-return pulse.* Suppose that the first pulse has a flip angle of 45 degrees and is applied in the x-direction in the rotating reference frame. Fat and water magnetizations immediately after the pulse are shown in Figure 6.6(a). If the interval between pulses is $1/2\delta\nu$, the second 45 degree pulse applied along the negative x-axis (Figure 6.6(b)) returns the water magnetization to the z-axis while placing the magnetization of fat into the transverse plane (Figure 6.6(c)).

Recently, several techniques for simultaneous spectral and spatial excitation have been introduced [14,15]. As in the Dixon method, magnetic field nonuniformity presents a problem for selective excitation of fat or water. As a result these techniques, as well as CHESS and excitation by binomial pulses, require sufficient magnetic field homogeneity in order to achieve good discrimination between water and fat.

Another MR approach for fat/water decomposition employs the difference in T_1 relaxation times of fat and water[2] in order to null the signal from fat while still retaining significant signal from water. In practice this approach is implemented by using inversion-recovery imaging. If T_1 of fat is indicated as T_{1fat}, then the inversion time needed to null the signal from fat is given by (see Chapter 2)

$$TI = T_{1fat} \ln \left[\frac{2}{1 + \exp(-TR/T_{1fat})} \right]. \qquad (6.2.2)$$

---

[2] The reported T_1 values for fat vary from 150 msec to 400 msec [16], while the typical T_1 of water-based tissues at 1.5 Tesla is greater than 500 msec.

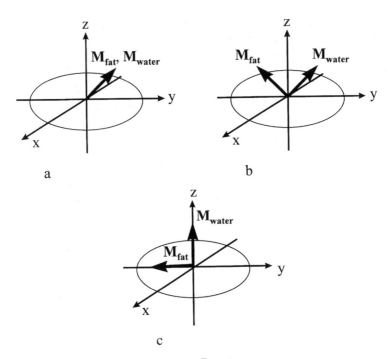

Figure 6.6. Excitation of fat by a 1–$\bar{1}$ pulse sequence: (a) fat and water magnetizations immediately after the first 45 degree pulse applied in the x-direction; (b) fat and water magnetizations immediately before the second 45 degree pulse applied along the negative x-axis; (c) immediately after the second pulse the water magnetization is along the z-axis while the fat magnetization is in the transverse plane.

Since T_{1fat} can vary depending on patient and anatomical site, the inversion time TI is often adjusted manually during the MR scan until the TI that causes the greatest decrease in the fat signal is found. Note that this approach is insensitive to the presence of magnetic field inhomogeneities because it does not rely on the fat/water chemical shift.

6.3. ARTIFACTS DUE TO MOTION AND FLOW

Because in conventional MRI *in vivo* the time required for spatially selective excitation, phase, and frequency encoding is normally about 100 msec or less, patient's motion during a single phase-encoding cycle can be considered minor. However, the time (a few minutes) needed to

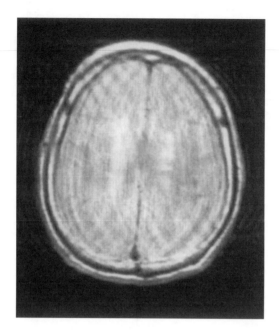

Figure 6.7. Spin-echo image of the head. The visible image blurring was caused by the patient motion during imaging.

acquire a complete set of 128 or 256 phase-encoded signals is long compared with the time scale for various physiologic motions such as respiration, cardiac motion, etc. As a result, motion-related changes in the specimen during different phase-encoding cycles create a potential source of artifacts in MR images.

Motion can perturb both amplitude and phase of the nuclear magnetization. Perturbations of the magnetization amplitude are caused by spins entering and exiting a voxel during MR imaging. A common result of such perturbations is image blurring, which can vary from very mild to very severe depending on the range and velocity of patient's motion (see Figure 6.7).

To examine motion-induced changes in the phase of the magnetization, let us consider for simplicity one-dimensional motion of spins in the direction of an applied gradient (e.g., G_x). The trajectory of a spin can be expanded in a series

$$x(t) = x(0) + v_x(0)t + \frac{a_x(0)t^2}{2} + \ldots, \qquad (6.3.1)$$

where $x(0)$, $v_x(0)$, and $a_x(0)$ are the position, velocity, and acceleration of the spin at $t = 0$, respectively. The motion-induced phase can be written as

$$\phi(t) = \gamma x(0) \int_0^t G_x \, dt' + \gamma v_x(0) \int_0^t G_x t' \, dt' + \frac{\gamma a_x(0)}{2} \int_0^t G_x t'^2 \, dt' \ldots$$

$$+ \frac{\gamma x^{(n)}(0)}{n!} \int_0^t G_x t'^n \, dt' + \ldots. \tag{6.3.2}$$

The first term on the right side of this equation describes the phase shift that occurs in the absence of motion. The second term is caused by motion with constant velocity. The third and higher order terms result from nonstationary motion. We will discuss two important artifacts resulting from the motion-dependent phase: image "ghosts" that occur in the presence of periodical motion of spins [17–19] and signal loss due to distribution of spin velocities in a voxel [20, 21].

Image Ghosts

One of the most common motion-induced artifacts is the appearance of image "ghosts," which normally occurs in the presence of periodical motion such as cardiac motion, respiratory motion, or pulsatile blood flow (see Figure 6.8). Regardless of the specific nature of the periodical motion, the resultant image artifacts appear as a series of ghost images of the moving structures. To understand the mechanism of these artifacts, we will consider a specific case when image ghosts are generated by pulsatile flow.

An analytical description of the ghosting artifact can be obtained by using a model [19] in which velocity of spins v_x during phase-encoding cycles is given by the sum of a constant term $v_{0,x}$ and a periodical modulation. That is, we assume

$$v_x = v_{0,x} + v_{1,x} \sin(2\pi m f_p TR + \theta), \tag{6.3.3}$$

where $v_{1,x}$ and f_p are the amplitude and frequency of modulation, respectively; TR is the repetition time; m is an integer that defines a phase-encoding cycle (see Chapter 3); θ is an arbitrary phase. Flow pulsations in the presence of the readout gradient G_x alter the phase of the magnetization according to the Eq. (6.3.2). Assuming that the excitation pulse is applied at time $t = 0$, the flow-induced phase shift

Figure 6.8. Gradient-echo images of a phantom composed of tubing with flowing liquid (water with added Gd-DTPA) immersed in a water bath. The phantom was connected to an electric pump. Image in (a) depicts stationary flow in the phantom. The artificial ghosts seen in (b) are caused by flow pulsations with frequency of 1 Hz.

of the gradient echo during the mth phase-encoding cycle is given by

$$\phi(m) = \phi_1 + \phi_2(m), \tag{6.3.4}$$

where $\phi_1 = \gamma M_{1,x} v_{0,x}$, $\phi_2(m) = \gamma M_{1,x} v_{1,x} \sin(2\pi m f_p TR + \theta)$; $M_{1,x} = \int_0^{TE} G_x t' \, dt'$ is the first order moment of the gradient waveform. Note that the second term on the right side of Eq. (6.3.4) describes the periodically varying phase induced by the motion of spins. We shall show that this periodicity causes image ghosts in the direction of the phase encoding.

Taking into account the motion-induced phase of spins, the signal in standard gradient-echo imaging can be written as

$$S(k_x, k_y(m)) = \iint M_{xy}(x, y) e^{jk_x + jk_y(m)y + j\phi_1 + j\phi_2(m)} \, dx \, dy. \tag{6.3.5}$$

The factor $\exp(j\phi_2(m))$ can be expanded in a series

$$e^{j\phi_2(m)} = \sum_{l=-\infty}^{\infty} J_l(\gamma M_{1,x} v_{1,x}) e^{j2\pi l m f_p TR + jl\theta}, \tag{6.3.6}$$

where J_l is a Bessel function of order l. Using the last two equations we obtain:

$$S(k_x, k_y(m)) = \iint \sum_{l=-\infty}^{\infty} J_l(\gamma M_{1,x} v_{1,x}) M_{xy}(x, y - l\delta y)$$
$$\times e^{jk_x x + jk_y(m)y + j\phi_1 + jl\theta} \, dx \, dy, \tag{6.3.7}$$

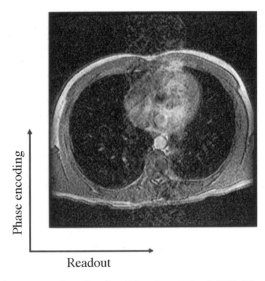

Figure 6.9. An example of pulsatile ghosts in MRI. The visible image ghosts in a gradient-echo image of the chest were produced by blood flow pulsations in the aorta.

where $\delta y = f_p TRL_y$, $L_y = 2\pi m/k_y(m)$ is the field-of-view in the phase-encoding direction. The resulting image of the structure with pulsatile flow contains ghosts described by the terms with nonzero l in the series

$$\sum_{l=-\infty}^{\infty} J_l(\gamma M_{1,x} v_{1,x}) M_{xy}(x, y - l\delta y) e^{j\phi_1 + jl\theta}.$$

An example of image ghosts induced by blood flow pulsations is shown in Figure 6.9. The distance between the ghosts, δy, is defined by the frequency of velocity modulation, repetition time, and field of view. The relative intensity of the lth ghost, $J_l(\gamma M_{1,x} v_{1,x}) e^{jl\theta}$, is a function of the amplitude of modulation and the first moment of the gradient waveform. Ghosts that are located outside of the field of view are folded back into the image because of aliasing; therefore, image ghosts cannot be removed by increasing the separation between the ghosts. If ghosts and other structures in images overlap, then the resultant image intensity becomes a sum of the complex intensities of the ghosts and the overlapping structures. Under the condition that ghost and background magnetizations are out of phase, the ghost can appear darker than its surroundings, whereas in other cases the in-phase condition ensures the appearance of bright ghosts in images [19].

Flow-Induced Signal Loss

As shown earlier, motion of spins in the presence of applied gradients modulates the phase of magnetization. Since normally there is a spread of spin velocities within a voxel, the resultant effect is a loss of signal from the voxel due to dephasing of spins. To model this effect we assume that moving spins have one-dimensional velocity distribution that is uniform between $-v_{max}$ and $+v_{max}$, where v_{max} is the maximum velocity of spins. In the presence of a gradient G_x applied in the direction of flow the signal from the moving spins is given by

$$S = \frac{\xi}{2v_{max}} \int_{-v_{max}}^{+v_{max}} \exp(j\gamma v \int_{0}^{t} G_x t' \, dt') \, dv = \frac{\xi \sin(\gamma M_{1,x} v_{max})}{\gamma M_{1,x} v_{max}}, \qquad (6.3.8)$$

where $M_{1,x} = \int_0^t G_x t' \, dt'$ and ξ is a constant that does not depend on velocity. Equation (6.3.8) shows that the flow induced signal loss is small when $|\gamma M_{1,x} v_{max}| \ll 1$. For $|\gamma M_{1,x} v_{max}| > 1$ the signal loss becomes significant. It is important to realize that signal loss depends on the velocity distribution. For example, one can show that a Gaussian velocity distribution leads to signal loss that is exponentially dependent on the square of the product of the first moment of the gradient waveform and the width of the velocity distribution.

6.4. MRI TECHNIQUES FOR MOTION ARTIFACT REDUCTION

Motion-induced artifacts such as blurring and image ghosts represent a major problem in diagnostic MRI because they impede visualization of small structures in images. In this section we discuss several techniques used to reduce motion artifacts in MRI. Because each of these techniques alone has its limitations, they can be used in combination to achieve better results.

Physical Restraint

Motion-induced artifacts can be reduced by decreasing the range of motion that causes the artifacts. In diagnostic MRI a reduction in patient motion is often achieved by using different immobilization devices. Suspended respiration is another popular technique used to eliminate respiratory motion of the patient within a breath-holding interval during imaging [22]. Due to its simplicity some form of physical restraint is frequently used during clinical MRI exams, although success of this approach ultimately depends on the patient's cooperation.

Signal Averaging

Signal averaging is the approach in which signal acquisition is repeated a number of times for each phase-encoding. The repeatedly acquired signals are subsequently averaged together in order to reduce the contribution from different physiological motions of the patient. The number of times the signals are averaged together will be referred to as *NSA*. Signal averaging can be performed by acquiring a series of data sets, each composed of *M* phase encodings. This approach is known as *serial averaging*. An alternative strategy, known as *parallel averaging*, is to repeat acquisitions for each phase encoding before changing the phase-encoding gradient. In the case of periodical motion, parallel averaging is ineffective when $NSA \times TR$ is much smaller than the period of motion because the averaged signals are acquired at the same stage of the motion cycle. Serial averaging is generally more effective for reducing image ghosts because it allows a longer interval between repetitive measurements thereby increasing the likelihood that the averaged signals are acquired at different stages of the motion cycle.

Averaging is a simple method for motion artifact reduction in MRI. Averaging does not require any special equipment and normally can be implemented with every pulse sequence. The major disadvantage of averaging is prolonged scan time, which is directly proportional to *NSA*. Another disadvantage is that averaging generally does not reduce motion-induced blurring in images.

Respiratory Gating

This approach is based on constant monitoring of patient position during respiratory motion by using a special transducer attached to the patient [23]. The obtained information is used either to trigger imaging pulse sequence when the patient is in a proper position or to reject a number of signals acquired during patient motion. Respiratory gating reduces ghosting and blurring artifacts because it minimizes motion-induced variations in the signal. The major disadvantage of respiratory gating is that it can double or triple scan time in diagnostic MRI. Scan time with respiratory gating can be reduced at the expense of increased blurring and ghosting artifacts if respiratory gating is implemented only during a fraction of data acquisition [24].

Cardiac Gating

Cardiac motion and related blood flow pulsations present a serious problem for *in vivo* MR imaging with scan times longer than the duration of the cardiac cycle. The basic idea of cardiac gating is to synchronize MR imaging with the cardiac cycle so that the same magnetization is imaged during successive phase encodings. In practice, cardiac gating is implemented by monitoring the electrocardiogram (ECG) signal produced during the cardiac cycle including contraction (systole) and relaxation (diastole). The ECG signal (a few mV) can be measured by applying electrodes to the body. Alternatively, cardiac motion can be monitored by recording an optical signal from a peripheral pulse of arterial blood (e.g., in a fingertip).

Detected ECG signal can be used to synchronize successive data acquisitions by repeatedly triggering a pulse sequence at a certain stage of the cardiac cycle. Consequently, in this approach the shortest repetition time, *TR*, is equal to the cardiac period. One disadvantage of ECG triggering is that it results in a relatively long *TR* of about 1 sec, which in turn leads to long scan times in diagnostic MRI. Another disadvantage is the appearance of artifacts caused by variations in *TR* due to irregular heart beats.

In an alternative approach, known as *retrospective gating*, MR imaging is performed continuously with fixed *TR*, which is much shorter than the period of the cardiac cycle [25]. The information about cardiac motion during imaging is constantly recorded. After completion of data acquisition, this information is used for interpolation and sorting of the data according to the corresponding phases of the cardiac cycle. Subsequently a number of images depicting different stages of cardiac motion are reconstructed. The major advantage of MR imaging with retrospective gating is its ability to depict the dynamics of the entire cardiac cycle.

Ordered Phase-Encoding Measurements

Ghosts that occur in MR images acquired in the presence of periodic motion are caused by periodic modulation of the magnetization during image acquisition. To eliminate image ghosts Perman *et al.* [19] suggested using a random order of phase encoding which, in the presence of periodic motion, leads to a random modulation pattern. The resulting effect of the randomly ordered phase encoding is a noise-like appearance of the motion artifact in contrast to the more conspicuous appearance of image ghosts in the case of a sequential (standard) order of phase encoding [19].

In another approach called ROPE (respiratory ordering of phase encoding), the phase-encoding order is defined by the patient position [26]. As in the case of respiratory gating, patient displacements in ROPE are monitored by a special transducer attached to the patient. The collected information is used to acquire spatial frequencies in the phase direction, k_y, at certain stages of the patient motion. In the case of periodical motion (e.g., in the phase-encoding direction), the displacement of the patient, y, becomes a periodical function of k_y if the standard order of phase encoding is used. This periodicity gives rise to image ghosts in the phase-encoding direction. The objective of ROPE is to make $y(k_y)$ a monotonic rather than a periodic function of k_y by choosing the sequence of phase encoding according to the measured patient motion. Although it can be difficult to implement in practice, ROPE significantly reduces image ghosts without increasing scan time (unlike averaging and respiratory gating).

Selective Saturation

In many instances motion-induced artifacts can be significantly reduced by selectively saturating moving structures which can be omitted from the images. A slab of tissue (oblique, parallel or perpendicular to the imaging plane) that contains such structures, can be saturated by applying a series of frequency selective r.f. pulses in the presence of the slice-select gradient. Because the magnetization of spins in the saturated slab is small, the moving structures in the slab do not generate significant artifacts in images. It should be noted that selective saturation is frequently used to improve differentiation between arteries and veins in MR images by saturating a slab (normally adjacent to the imaged volume) that contains these blood vessels. Since the blood entering the image volume through the slab has a low magnetization, it appears dark in images and causes minimal image distortion due to blood flow pulsations (Figure 6.10).

Gradient Moment Nulling

Motion induced phase shifts are linearly dependent on the moments of the gradient waveform:

$$\mathbf{M}_n = \int_0^t t'^n \mathbf{G}\, dt',$$

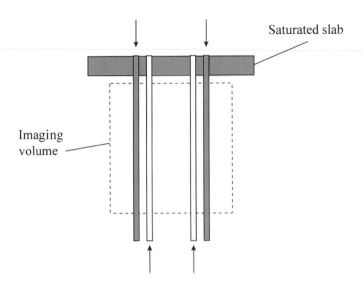

Figure 6.10. Schematic of selective saturation of flow. The arrows indicate the direction of flow. The flow entering the imaged volume from above has a low magnetization; therefore, it appears darker than the unsaturated flow entering the image volume from below.

$n = 1, 2, \ldots$ The gradient moment nulling technique suppresses motion artifacts by using specially designed gradient waveforms with $\mathbf{M}_n = 0$ [27–29]. For example, if the gradient pulses have zero $\mathbf{M}_1$ and $\mathbf{M}_2$, then the accumulated phase no longer depends on velocity or acceleration. Although gradient waveforms can be designed to null any chosen number of moments, nulling of higher-order moments requires a number of additional gradient lobes that significantly increase the minimum TE of a pulse sequence. As a result, gradient moment nulling is most often implemented by using gradient waveforms that only null the first-order moment, $\mathbf{M}_1$ (Figure 6.11).

Fractional Echo

Since motion-dependent phase shifts occur only in the presence of magnetic field gradients, they can be reduced by shortening the duration of the gradient waveforms. To implement a pulse sequence with a compact readout gradient, Nishimura *et al.* [30] suggested using a bipolar gradient with a very short dephasing lobe applied prior to data acquisition. As a result, during the signal acquisition, only a fraction of the gradient echo is collected (Figure 6.12). Although fractional echo

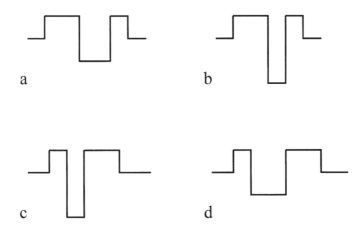

Figure 6.11. Examples of gradient waveforms with first order moment nulling used for slice selection (a and b) and readout (c and d).

acquisition tends to decrease SNR due to decreased readout time, the reduced dephasing of spins can actually increase the resultant SNR.

In the fractional-echo approach, the acquired spatial frequencies are used to synthesize the missing part of the data. The reconstruction algorithms can be understood by considering an ideal case when image intensity is described by a real function. In this case, the k-space data,

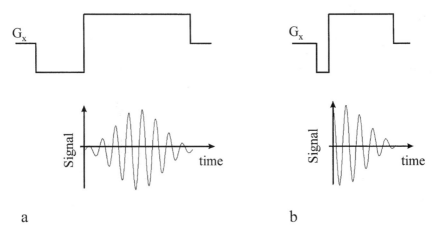

Figure 6.12. Standard (a) and fractional echo (b) acquisitions in gradient-echo imaging.

S, are Hermitian:

$$S(-k_x, -k_y) = S^*(k_x, k_y), \qquad (6.4.1)$$

where $S^*(k_x, k_y)$ is the complex conjugate of $S(k_x, k_y)$. From Eq. (6.4.1), it follows that acquisition of only half of spatial frequencies in the readout direction makes it possible to synthesize the unsampled half of the data. In practice magnetic field inhomogeneities, motion-induced phase shifts and other factors give rise to complex image intensity characterized by a spatially varying phase. Image reconstruction from partially acquired data in the latter case is discussed in Chapter 7.

Note that compared with conventional gradient-echo acquisition, fractional-echo acquisition can be implemented with shorter echo time. This makes fractional-echo acquisition a useful technique for rapid gradient-echo imaging performed with very short *TE* and *TR* (see Chapter 7).

Navigator Echoes

The navigator approach is based on the assumption of a global linear motion of the imaged object. In this approach, object motion is recorded by acquiring special "navigator" echoes [31,32]. Motion-induced phase shifts of these echoes are subsequently used for phase correction of the acquired signals.

To describe this approach we consider, for simplicity, the case of one-dimensional motion in the readout direction. In this case the acquired signal in 2D imaging can be expressed as

$$S(k_x, k_y(m)) = \iint M_{xy}(x - \Delta x(m), y)e^{jk_x x + jk_y(m)y}\, dx\, dy. \qquad (6.4.2)$$

In this equation the readout and phase-encoding gradients are in the x- and y-directions, respectively; $\Delta x(m)$ is the displacement at the time of a mth phase encoding. Equation (6.4.2) can be written as

$$S(k_x, k_y(m)) = e^{jk_x \Delta x(m)} \iint M_{xy}(x, y)e^{jk_x x + jk_y(m)y}\, dx\, dy$$

$$= e^{jk_x \Delta x(m)} S_0(k_x, k_y(m)), \qquad (6.4.3)$$

where $S_0(k_x, k_y(m))$ is the signal from the same object at rest. From this equation it follows that the signal from a linearly moving object equals the signal from the object at rest multiplied by a phase factor $e^{jk_x \Delta x(m)}$. The phase factor can be determined from an additional "navigator" echo acquired before applying the phase-encoding gradient

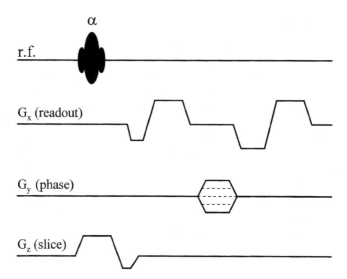

Figure 6.13. Gradient-echo pulse sequence with acquisition of a navigator echo in the readout direction.

(Figure 6.13). The navigator signal, $S_{nav,m}(k_x)$, acquired during the mth phase encoding is given by

$$S_{nav,m}(k_x) = e^{jk_x \Delta x(m)} S_{nav,0}(k_x), \qquad (6.4.4)$$

where $S_{nav,0}(k_x)$ is the reference navigator signal (e.g., acquired before the first phase encoding). By using navigator echoes, the signal from the object at rest can be restored as follows:

$$S_0(k_x, k_y(m)) = \frac{S_{nav,0}(k_x)}{S_{nav,m}(k_x)} S(k_x, k_y(m)). \qquad (6.4.5)$$

This equation demonstrates that navigator signals can be used to eliminate the phase factor caused by motion in the direction of the navigator's gradient. In general, three orthogonal navigator gradients are required for phase correction in the case of three-dimensional linear motion.

REFERENCES

[1] J.M.S. Hutchison, R.J. Sutherland, J.R. Mallard. "NMR imaging: image recovery under magnetic field with large nonuniformities," *J. Phys. E: Sci. Instrum.*, **11**, 217 (1978).

[2] K.M. Ludeke, P. Roschmann, R. Tischler. "Susceptibility artifacts in NMR imaging," *Magn. Reson. Imaging*, **3**, 329 (1985).

[3] K. Sekihara, S. Matsui, H. Kohno. "NMR imaging for magnets with large nonuniformities," *IEEE Trans. Med. Imag.*, **MI–4**, 193 (1985).

[4] F. Farzaneh, S.J. Riederer, N.J. Pelc. "Analysis of T2 limitations and off-resonance effects on spatial resolution and artifacts in echo-planar imaging," *Magn. Reson. Med.*, **14**, 123 (1990).

[5] J. Frahm, K.D. Merboldt, W. Hanicke. "The effects of intravoxel dephasing and incomplete slice refocusing on susceptibility contrast in gradient-echo MRI," *J. Magn. Reson. B.*, **109**, 234 (1995).

[6] J.R. Reichenbax, R. Venkatesan, D.A. Yablonskiy, M.R. Thompson, S.Lai, E.M. Haacke. "Theory and applcation of static field inhomogeneity effects in gradient-echo imaging," *JMRI*, **7**, 266 (1997).

[7] B. Romner, M. Olsson, B. Ljunggren, S. Holtas, H. Saveland, L. Brandt, B. Persson. "Magnetic resonance imaging and aneurysm clips," *J. Neurosurg.*, **70**, 426 (1989).

[8] W.T. Dixon. "Simple proton spectroscopic imaging," *Radiology*, **153**, 189 (1984).

[9] J. Szumowski, D.B. Plewes. "Separation of lipid and water MR imaging signals by chopper averaging in the time domain," *Radiology*, **165**, 247 (1987).

[10] G.H. Glover, E. Schneider. "Three-point Dixon technique for true water/fat decomposition with B_0 inhomogeneity correction," *Magn. Reson. Med.*, **18**, 371 (1991).

[11] A. Haase, J.Frahm, W. Hanicke, D. Matthaei. "^{1}H NMR chemical shift selective (CHESS) imaging," *Phys. Med. Biol.*, **30**, 341 (1985).

[12] V. Sklenar, Z. Starcuk. "1-2-1 pulse train: a new effective method of selective excitation for proton NMR in water," *J. Magn. Reson.*, **50**, 495 (1982).

[13] P.J. Hore. "Solvent suppression in Fourier transform nuclear magnetic resonance," *J. Magn. Reson.*, **55**, 283 (1983).

[14] C.H. Meyer, J.M. Pauly, A. Macovski, D.G. Nishimura. "Simultaneous spatial and spectral selective excitation," *Magn. Reson. Med.*, **15**, 287 (1990).

[15] F. Schick. "Simultaneous highly selective MR water and fat imaging using a simple new type of spectral-spatial excitation," *Magn. Reson. Med.*, **40**, 194 (1998).

[16] J. Szumowski, J.H. Simon. "Proton chemical shift imaging," *Magnetic Resonance Imaging*. D.D. Start and W.G. Bradly, eds., Mosby-Year Book, 1992.

[17] C.L. Schultz, R.J. Alfidi, A.D. Nelson, S.Y. Kopiwoda, M.E. Clampitt. "The effect of motion on two-dimensional Fourier transformation magnetic resonance images," *Radiology*, **152**, 117 (1984).

[18] M.L. Wood, R.M. Henkelman. "MR image artifacts from periodic motion," *Med. Phys.*, **12**, 143 (1985).

[19] W.H. Perman, P.R. Moran, R.A. Moran, M.A. Bernstein. "Artifacts from pulsatile flow in MR imaging," *J. Comput. Assist. Tomogr.*, **10**, 473 (1986).

[20] G.L. Naylor, D.N. Firmin, D.B. Longmore. "Blood flow imaging by cine magnetic resonance," *J. Comp. Assist. Tomogr.*, **10**, 715 (1986).

[21] C.B. Ahn, S.Y. Lee, O. Nalcioglu, Z.H. Cho. "The effects of random directional distributed flow in nuclear magnetic resonance imaging," *Med. Phys.*, **14**, 43 (1987).

[22] R.R. Edelman, P.F. Hahn, R. Buxton, J. Wittenberg, J.T. Ferrucci, S. Saini, T.J. Brady. "Rapid MR imaging with suspended respiration: clinical application in the liver," *Radiology*, **161**, 125 (1986).

[23] V.M. Runge, J. A. Clanton, C.L. Partain, A.E. James. "Respiratory gating in magnetic resonance imaging at 0.5 Tesla," *Radiology*, **151**, 521 (1984).

[24] M.L. Wood, R.M. Henkelman. "Suppression of respiratory motion artifacts in magnetic resonance imaging," *Med. Phys.*, **13**, 794 (1986).

[25] G.H. Glover, N.J. Pelc. "A rapid-gated cine MRI technique," *Magn. Reson. Annual*, 299 (1988).

[26] D.R. Bailes, D.J. Gilderable, G.M. Bydder, A.G. Collins, D.N. Firmin. "Respiratory ordering of phase encoding (ROPE): a method for reducing motion artifacts in MR imaging," *J. Comp. Assist. Tomogr.*, **9**, 835 (1985).

[27] P.M. Pattany, J.J. Phillips, L.C. Chiu, J.D. Lipcamon, J.L. Duerk, J.M. McNally, S.N. Mohapatra. "Motion artifact suppression technique (MAST) for MR imaging," *J. Comput. Assist. Tomogr.*, **11**, 369 (1987).

[28] E.M. Haacke, G.W. Lenz. "Improving image quality in the presence of motion by using rephasing gradients," *AJR*, **148**, 1251 (1987).

[29] J.G. Pipe, T.L. Chenevert. "A progressive gradient moment nulling design technique," *Magn. Reson. Med.*, **19**, 175 (1991).

[30] D.G. Nishimura, A. Macovski, J.I. Jackson, R.S. Hu, C.A. Stevick, L. Axel. "Magnetic resonance angiography by selective inversion recovery imaging using compact gradient-echo sequence," *Magn. Reson. Med.*, **8**, 96 (1988).

[31] R.L. Ehman, J.P. Felmlee. "Adaptive technique for high-definition MR imaging of moving structures," *Radiology*, **173**, 255 (1989).

[32] H.W. Korin, F. Farzaneh, R.C. Wright, S.J. Riederer. "Compensation for effects of linear motion in MR imaging," *Magn. Reson. Med.*, **12**, 99 (1989).

CHAPTER 7

Rapid MR Imaging

The need to reduce motion artifacts and overall scan time in diagnostic MRI provided the initial impetus for the development of rapid MR imaging. Rapid MRI with scan times of 0.1–1 sec per image has found a number of important clinical applications such as functional, perfusion, and cardiac imaging. Because of its vast potential and already proven importance for biomedical imaging, rapid MR scanning has been an active area of MRI research.

A reduction in MRI scan time is commonly achieved by using two different approaches individually or in combination. The underlying strategy in the first approach is to shorten the time needed to collect NMR signals required for image reconstruction. This can be achieved by using a number of developed techniques such as gradient-echo imaging with short repetition times [1–3], echo-planar imaging [4,5], fast spin-echo imaging [6], and spiral imaging [7,8]. These imaging techniques will be discussed in more detail in the following sections of this chapter. In the second approach, known as *partial k-space acquisition*, a reduction in scan time is achieved by collecting only a fraction of full k-space data required to implement the conventional reconstruction algorithm. Image reconstruction in this approach is performed by using additional assumptions about the properties of the imaged object. Two of the partial k-space acquisition techniques will be discussed in the last section of this chapter.

7.1. RAPID GRADIENT-ECHO IMAGING

Scan time in conventional MR imaging is given by the product of repetition time, TR, total number of phase encodings, and number of times the signal is averaged, NSA. Assuming that $NSA = 1$, only two options exist for reducing scan time in the conventional approach: decreasing the number of phase encodings and shortening TR. Because in practice the minimum number of phase encodings is limited by the required spatial resolution, short scan times in rapid MR imaging with standard frequency and phase encodings are achieved by using very short TR (normally on the order of 10 msec).

Techniques for rapid gradient-echo imaging can be divided into two different groups. The first group is composed of pulse sequences that employ r.f. and gradient spoiling to destroy the transverse magnetization after each signal acquisition while maintaining the steady-state longitudinal magnetization. The second group consists of pulse sequences with gradient refocusing used to sustain both transverse and longitudinal components of the magnetization in equilibrium.

"Spoiled" Gradient-Echo Imaging

Successful implementation of conventional MR imaging requires the existence of a steady-state in which the same magnetization in the object is created after each r.f. excitation. It has been shown [9–11] that a steady-state can be established by a train of repetitive r.f. pulses even when the local magnetic field is nonuniform, or when identical external gradients are applied between successive excitations. On the other hand, in conventional MR imaging the phase-encoding gradient is varied from one phase-encoding cycle to the next. It is therefore clear that the phase of the transverse magnetization also changes from cycle to cycle. The resultant dynamics of the magnetization during imaging critically depends on the interval between successive excitations. If TR is much longer than T_2, then the transverse magnetization is negligibly small prior to each excitation. In this case a steady-state exists despite the fact that the phase of the transverse magnetization varies from excitation to excitation. On the other hand, when TR is smaller than T_2 the variations in the phase of magnetization do not allow the existence of a steady-state.

To reduce nonequilibrium effects, the transverse magnetization in the sample can be destroyed (spoiled) by additional external gradients applied after every signal acquisition. These gradients (also known as *crusher gradients*) cause dephasing of spins, which in turn reduces the average transverse magnetization in a voxel. The most effective spoiling of the transverse magnetization is achieved when in addition to gradient spoiling the phase of r.f. pulses is varied from excitation to excitation [11,12]. A basic 2D gradient-echo sequence, known

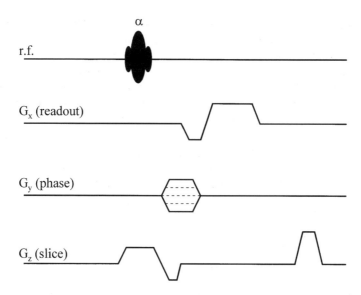

Figure 7.1. 2D FLASH sequence with a spoiling gradient applied in the z-direction.

as FLASH (fast low-angle shot) or spoiled GRASS (gradient recalled acquisition in the steady-state), with gradient "spoiling" is shown in Figure 7.1. An example of a FLASH image is shown in Figure 7.2.

Gradient-echo image intensity in FLASH can be approximated as

$$I \propto N(H) \exp(-TE/T_2^*) \sin \alpha \frac{1 - E_1}{1 - E_1 \cos \alpha}, \qquad (7.1.1)$$

where $N(H)$ is the proton density, $E_1 = \exp(-TR/T_1)$, and α is the flip angle. When imaging with TR significantly longer than T_1 the longitudinal magnetization almost completely recovers during the interval between successive excitations. In this case 90 degree excitation pulses provide the greatest signal. Conversely, when repetition time is much shorter than T_1 the magnetization does not have enough time for recovery. In the latter case, flip angles smaller than 90 degrees should be used to maximize the signal (see Chapter 2).

It should be noted that decreasing TR results in SNR penalty. The maximum signal, achieved when imaging with the Ernst flip angle (i.e., $\cos \alpha = E_1$), varies as $\sqrt{TR/T_1}$ when $TR \ll T_1$. Therefore, we can conclude that at short TR the SNR is proportional to $\sqrt{TR}$. However, this reasoning does not take into account the fact that imaging with

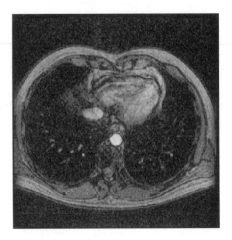

Figure 7.2. 256 (readout) × 160 (phase encoding) FLASH image of a transverse section of the chest. Imaging parameters: field-of-view = 34 cm × 34 cm, slice thickness = 7 mm, $TR = 40$ msec, $TE = 2.2$ msec, flip angle = 40 degrees.

ultrashort TR (a few milliseconds) requires the use of very short readout time, which further reduces the SNR.

Equation (7.1.1) shows that in the case when $TR \ll T_2^*$, image contrast is defined by T_1 and proton density. The lack of T_2 contrast and low SNR are the main limitations of short TR spoiled gradient-echo imaging. To address the issue of T_2-dependent contrast, Haase [3] has suggested using a modified FLASH pulse sequence in which nuclear magnetization in a sample is specially prepared prior to imaging. A magnetization-prepared pulse sequence used to acquire T_2-weighted images contains an additional sequence of 90, 180, and 90 degree pulses (Figure 7.3). After the last 90 degree pulse flips magnetization back to the z-axis, the longitudinal magnetization is given by

$$M_z = M_0 \exp(-\tau/T_2), \qquad (7.1.2)$$

where τ is the interval between the first and the last preparation pulses. Crusher gradients are then used to destroy any residual transverse magnetization that exists due to incomplete refocusing of spins following the 180 degree pulse. After the preparation period, "snapshot" FLASH imaging is performed with low flip angle ($\alpha < 5°$) and short repetition times (a few milliseconds) in order to decrease effects caused by T_1 relaxation. Other modifications of FLASH allow acquisition of inversion-recovery and diffusion-weighted images [3, 13]. FLASH

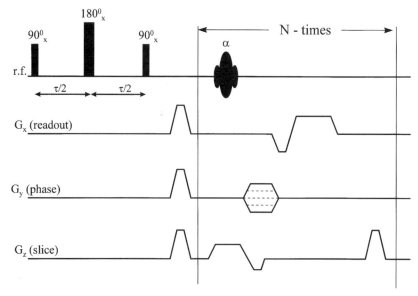

Figure 7.3. Magnetization-prepared snapshot FLASH sequence for acquiring T_2-weighted images.

imaging of the prepared magnetization is frequently referred to as MP-RAGE (magnetization prepared rapid acquisition gradient-echo).

Refocused Gradient-Echo Imaging

The idea of this approach, frequently referred to as FISP (fast imaging with steady-state precession) or SSFP (steady state free precession), is to create conditions under which both longitudinal and transverse magnetization are in equilibrium [14, 15]. The problem of accumulated phase changing from excitation to excitation due to the varying phase-encoding gradient is solved by applying a gradient (sometimes called rewinding gradient) with reverse polarity in the end of each repetition period. In principle, only refocusing along the phase-encoding direction is necessary because the slice-select and frequency-encoding gradients do not vary from cycle to cycle. However, in order to null the accumulated phase rewinding gradients can be applied along all axes.

Hinshow [16] has shown that in the case when the phase of r.f. pulses alternates from cycle to cycle, the steady-state magnetization can be relatively large even in the case of extremely short TR. Neglecting magnetic field inhomogeneities, FISP image intensity obtained with alternate-phase excitations is given by

$$I \propto M_0\, e^{-TE/T_2} \frac{(1 - E_1)\sin\alpha}{1 - (E_1 - E_2)\cos\alpha - E_1 E_2}, \qquad (7.1.3)$$

where $E_2 = \exp(-TR/T_2)$ [16]. From this equation it follows that contrast in FISP images acquired with long repetition time (i.e., $TR > T_2$) is similar to contrast in spoiled gradient-echo imaging. At very short repetition time (i.e., $TR \ll T_2, T_1$) we obtain from Eq. (7.1.3)

$$I \propto \frac{M_0 \sin\alpha}{1 + T_1/T_2 - (T_1/T_2 - 1)\cos\alpha}. \qquad (7.1.4)$$

One important conclusion drawn from this equation is that the signal does not depend on repetition time. Therefore, FISP imaging with very short TR results in much smaller SNR penalty as compared to FLASH. Another important result is that image contrast in FISP depends on the ratio of T_2/T_1. For example, when imaging with $\alpha = 90°$ image intensity varies as $(1 + T_1/T_2)^{-1}$. This explains why tissues with large T_2/T_1 ratios appear very bright in FISP images acquired with large flip angles.

7.2. ECHO-PLANAR IMAGING

Echo-planar imaging (EPI), originally suggested by Mansfield [4, 5], is one of the fastest MRI techniques currently used. A basic EPI pulse sequence is shown in Figure 7.4(a). Similar to conventional MR imaging, the signal acquisition in EPI is performed along parallel lines in k-space (Figure 7.4(b)). However, unlike conventional imaging EPI utilizes a very efficient sampling scheme by acquiring a number of lines in k-space after each excitation pulse. As a result EPI images can be produced with only a few excitations of the magnetization in the object. The fastest EPI scanning, known as *single-shot* EPI, is achieved when all required signals are collected after a single excitation pulse. Next we consider several EPI pulse sequences used in MR imaging.

Gradient-Echo EPI (GRE-EPI)

In GRE-EPI, a number (e.g., 128 or 256) of equidistant gradient echoes (echo train) with opposite polarities are produced by reversals of the frequency-encoding gradient (Figure 7.4(a)). The formation of an echo train in EPI can be understood by considering the phase of the transverse magnetization accumulated in the presence of the

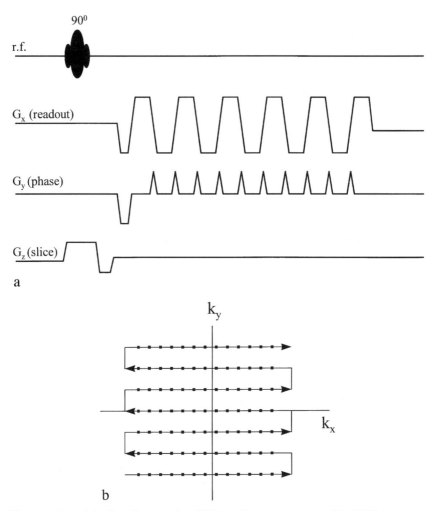

Figure 7.4. (a) Gradient-echo EPI pulse sequence; (b) EPI k-space sampling (arrows indicate the direction of sampling).

time-varying readout gradient. The peak of each gradient echo occurs in the center of the corresponding acquisition interval when the accumulated phase approaches zero (Figure 7.5). The echo-train time (ETT) is given by the product of the time interval between consecutive gradient echoes, referred to as *echo spacing* (ESP), and the number of gradient echoes generated after an excitation pulse.

Each line in the GRE-EPI k-space corresponds to a phase-encoded gradient echo. Phase encoding in EPI is normally implemented by

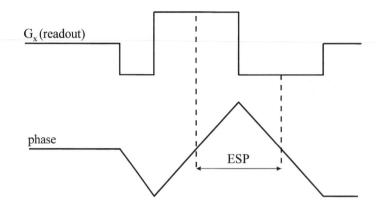

Figure 7.5. Diagram of the readout gradient reversals and the corresponding phase reversals in EPI.

applying "blips" of the phase-encoding gradient between reversals of the frequency-encoding gradient (Figure 7.4(a)). Each phase-encoding blip defines a new line in k-space. In contrast to the standard acquisition scheme, successive lines in the EPI k-space are sampled in opposite directions (Figure 7.4(b)).

The major drawback of the highly efficient k-space sampling in GRE-EPI is frequent image artifacts caused by T_2^* decay or fat/water chemical shift. Image artifacts can be particularly severe in the case of single-shot EPI. Consider, for example, EPI with the following parameters: 256(readout) × 128(phase encoding) imaging matrix, 256 kHz (±128 kHz) sampling bandwidth. Given these parameters each line in k-space is acquired during $256 \times 10^{-3}/256 \sec = 1 \text{msec}$. Assuming that the gradient ramp time (i.e., the time required to change the gradient from zero to its maximum amplitude) is much smaller than the acquisition time, we obtain ESP ≈ 1 msec and ETT ≈ 128 msec. Because the calculated ETT for single-shot EPI is significantly greater than characteristic T_2^* values in tissues (30–60 msec), gradient echoes at the tail of the echo train would be heavily T_2^*-weighted in contrast to gradient echoes in the beginning of the train. Suppose, for example, that phase encoding in EPI is implemented in such a way that low spatial frequency components of the signal are acquired at the beginning of the echo-train when the effect of T_2^* decay is relatively small, and high spatial frequencies are acquired at the tail of the echo-train when the signal significantly decreases due to T_2^* decay. The reduction in the amplitudes of high frequency components amounts to low-pass filtering of the signal resulting in decreased spatial resolution and increased blurring in EPI images.

Because of long readout time required to generate a number of gradient echoes after each excitation pulse, EPI images can display significant geometric distortion caused by magnetic field inhomogeneities and fat/water chemical shift. The potential severity of image distortion can be assessed by considering the effect of fat/water chemical shift in single-shot EPI. The phase difference between the water and fat magnetizations accumulated between consecutive gradient echoes is $2\pi\text{ESP}(\nu_w - \nu_f)$, where ν_w and ν_f are the resonance frequencies of water and fat, respectively. The accumulated phase causes spatial shift (misplacement) of fat in the direction of the phase encoding rather than in the readout direction; the converse is true in the case of conventional imaging. The magnitude of the fat misplacement in single-shot EPI images is $L_y\text{ESP}(\nu_w - \nu_f)$, where L_y is the field of view in the phase-encoding direction. At a field strength of 1.5 Tesla, the misplacement of fat in images acquired with 128 phase encodings and ESP of 1 msec is approximately 28 pixels. Because of such significant misplacement of fat, EPI is typically implemented with suppression of the fat signal.

Shortening of readout time is desirable because it decreases EPI image distortions. Readout time can be reduced by decreasing echo spacing, decreasing the number of phase encodings at the expense of spatial resolution and increasing the number of excitations (referred to as shots) at the expense of increased scan time. Shortening echo spacing can be achieved by either increasing sampling bandwidth, which requires strong gradients and causes a decrease in SNR, or by reducing the number of readout points at the expense of decreased spatial resolution in the readout direction. In practice, single-shot EPI with high spatial resolution is performed by using specially designed gradient systems providing both high gradient strength and short ramp time (about 100 microseconds). It should be noted that compared with rapid gradient-echo imaging considered earlier, single-shot GRE-EPI generally allows shorter scan time (i.e., ~100 msec per image) due to its higher sampling efficiency and higher SNR. These advantages of gradient-echo EPI have proven extremely useful for functional imaging of neuron activation in the brain (see Chapter 4).

Because it uses a number of excitations to acquire image data, multi-shot EPI has shorter readout time compared with single-shot EPI. As a result, multi-shot EPI images display smaller distortion and loss of signal than single-shot EPI images. On the other hand, increased scan time makes multi-shot EPI more susceptible to motion artifacts than single-shot EPI.

Spin-Echo EPI

Spin-echo EPI (SE-EPI) is implemented by using a refocusing 180° pulse after spatially selective excitation. A typical SE-EPI pulse-sequence is shown in Figure 7.6(a). By choosing the appropriate ESP and phase-encoding order, the center of k-space is normally sampled at the peak of the spin echo. Like in conventional spin-echo imaging, contrast in SE-EPI images depends on the echo time, TE, which defines T_2-weighting of the center of k-space. Experience shows that due to refocusing of spins SE-EPI images are less sensitive to the presence of magnetic field inhomogeneities than GRE-EPI images. Compared with conventional spin-echo imaging, SE-EPI generally requires a significantly longer readout time. Consequently, SE-EPI is more susceptible to artifacts caused by different T_2 weighting of the acquired signals and to artifacts caused by magnetic field inhomogeneities.

Diffusion-Weighted EPI

Diffusion-weighted EPI (DW-EPI) is performed with an additional pair of strong pulsed gradients applied on either side of a 180 degree pulse (Figure 7.6(b)). The attenuation of image intensity due to diffusion is given by the factor $\exp(-bD)$, where b depends on the gradient strength and duration, and the time interval between the diffusion gradients, and D is the diffusion coefficient [17]. Since EPI employs a significant number of imaging gradients, the equation for b should also include the contribution from these gradients. The diffusion coefficient can be calculated from a series of images acquired with different values of b (but the same T_1 and T_2-weightings). Single-shot DW-EPI has found a number of applications in neuroimaging because DW-EPI is fast (~100 msec per image), and therefore, almost free from motion-induced artifacts that often degrade conventional diffusion-weighted images.

7.3. FAST SPIN-ECHO IMAGING

Fast spin-echo imaging proposed by Hanning *et al.* [6] employs a number of 180 degree pulses to produce a train of spin echoes after each excitation of the transverse magnetization. To understand the process of formation of a spin-echo train, consider a simple model

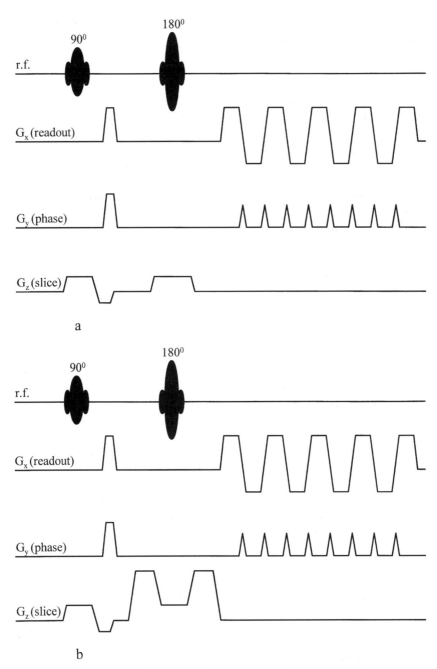

Figure 7.6. (a) SE-EPI sequence; (b) DWI-EPI sequence.

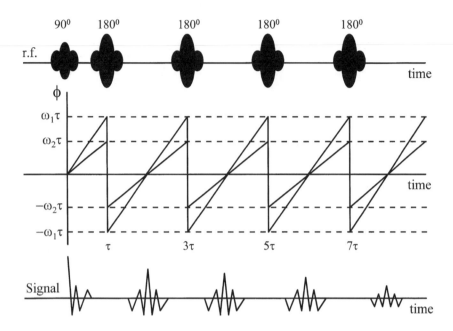

Figure 7.7. Formation of a spin-echo train in the presence of 180 degree
pulses.

system composed of two noninteracting spins. Suppose that these spins
have different Larmor frequencies ω_1 and ω_2. After a 90 degree excitation pulse, the phase of each spin will increase linearly with time:
$\phi_1 = \omega_1 t$, $\phi_2 = \omega_2 t$ (Figure 7.7). A first 180° pulse, applied at time τ
after the excitation, inverts the phases of spins. Following the pulse
a spin-echo is produced at time 2τ when the phases ϕ_1 and ϕ_2 approach
zero. Successive 180 degree pulses applied at $t = \tau$, $3\tau, 5\tau, 7\tau, \ldots$,
$(2N - 1)\tau$ cause a series of spin echoes at $t = 2\tau$, $4\tau, 6\tau, 8\tau, \ldots, 2N\tau$
(Figure 7.7). Spatial information can be encoded in the spin echoes
by applying frequency and phase-encoding gradients between successive 180 degree pulses. A basic fast spin-echo (FSE) pulse sequence,
also referred to as RARE (rapid acquisition with relaxation enhancement) or turbo spin-echo, is shown in Figure 7.8.

Like EPI, fast spin-echo imaging utilizes a very efficient sampling
scheme by acquiring several lines in k-space after a single excitation.
Because each line in the FSE k-space corresponds to a spin echo,
FSE acquisition reduces artifacts due to magnetic field inhomogeneities and fat/water chemical shift as compared to EPI. Because

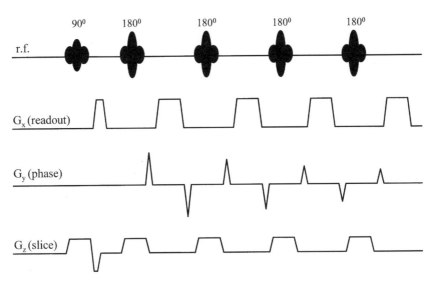

Figure 7.8. Diagram of a basic FSE sequence.

gradient echoes can be produced more rapidly than spin echoes, the echo spacing in EPI is usually several times shorter than the interval between successive spin-echoes (about 5–15 msec) during FSE acquisition. The relatively long echo spacing in FSE acquisition gives rise to image artifacts caused by different T_2-weighting of the acquired signals [18, 19]. T_2 decay also limits the number of echoes that can be acquired during single-shot FSE imaging. For example, echo-train length in a single-shot FSE acquisition with 128 phase encodings and echo spacing of 10 msec would be 1280 msec. Since typical T_2 relaxation time in tissue is about 100 msec, the spin echoes at the tail of the echo train would produce very little signal. This example can explain why high resolution FSE images are usually obtained by using multishot acquisition of 4–16 successive spin echoes produced after each excitation of the magnetization in a sample.

Contrast in fast spin-echo imaging is defined by the T_2-weighting of central lines in k-space corresponding to small values of the phase-encoding gradient. A parameter defining T_2-weighting of FSE images is the time interval, known as the *effective echo time*, between an excitation pulse and the center of a spin-echo with zero phase encoding. For example, heavily T_2-weighted FSE images are acquired with effective echo time equal to or greater than T_2 in tissue. In contrast, reduced T_2-weighting is achieved with effective echo time significantly shorter than T_2 (Figure 7.9).

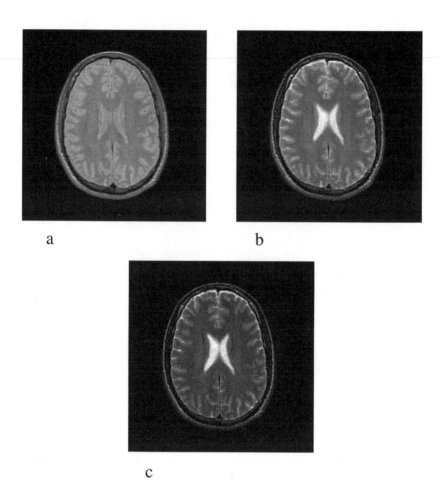

a   b

c

Figure 7.9. Dependence of T_2 contrast in FSE imaging on effective echo time: (a) $TE = 15\,\text{msec}$, (b) $TE = 75\,\text{msec}$, (c) $TE = 115\,\text{msec}$. FSE images of the brain were acquired with eight spin echoes per excitation using the following imaging parameters: matrix size $= 256(\text{readout})\times 192(\text{phase encoding})$, field-of-view $= 24\,\text{cm} \times 24\,\text{cm}$, slice thickness $= 2\,\text{mm}$, $TR = 3\,\text{sec}$.

FSE imaging has become widely used clinically because it pro-vides the required contrast between different tissues while being significantly faster than conventional spin-echo imaging. Contrast in FSE images can be enhanced further by using appropriately prepared magnetization. For example, in inversion-recovery (IR) FSE imaging the magnetization in the object is inverted first by a 180° pulse. Following the inversion, the magnetization experiences T_1 relaxation during a chosen delay after which FSE acquisition begins. Similar to conventional inversion-recovery imaging, IR FSE is frequently used to improve image contrast by nulling signals from certain tissues such as CSF.

7.4. SPIRAL IMAGING

Spiral imaging first demonstrated by Ahn *et al.* [7] is a very effective approach for rapid MRI. In contrast to EPI, spiral imaging utilizes smooth k-space trajectories that can be implemented on MR scanner without special hardware. Spiral acquisition reduces artifacts due to flow and motion as compared to conventional imaging with Cartesian k-space trajectories mainly because the center of k-space is acquired immediately after excitation with very little time for phase perturbations. Because of these attractive features, spiral imaging has been used for MR angiography, cardiac, and functional imaging.

In the following discussion, we restrict ourselves to the case of 2D spiral trajectories that can be described by using polar coordinates (k, ϕ). The relationship between Cartesian (k_x, k_y) and polar coordinates is given by

$$k_x = k \cos \phi, \qquad k_y = k \sin \phi. \tag{7.4.1}$$

Next we consider Archimedean spirals, which provide uniform radial sampling of k-space. The Archimedean spirals are described by the following equation [20]:

$$k(t) = A\phi(t), \tag{7.4.2}$$

where A is a constant. The radius of the acquired circular region in k-space, k_{max}, is defined by the required spatial resolution, Δr: $k_{max} = \pi/\Delta r$. If a spiral trajectory reaches k_{max} after N_{rev} revolutions, then the maximum polar angle is $\phi_{max} = 2\pi N_{rev}$ and from Eq. (7.4.2) it follows that

$$A = \frac{1}{2N_{rev}\Delta r}. \tag{7.4.3}$$

To avoid aliasing when imaging with $L \times L$ field-of-view, the radial distance between acquired k-space points cannot exceed $2\pi/L$. From this condition and Equations (7.4.2) and (7.4.3), it follows that spiral imaging with $N \times N$ matrix and M interleaves requires at least $N/(2M)$ revolutions for each of the spiral trajectories.

The Cartesian components of the gradient, $\mathbf{G}$, and components of the slew rate, $d\mathbf{G}/dt$, are defined by the time derivatives of the vector $\mathbf{k}$. That is,

$$
\begin{aligned}
G_x &= \gamma^{-1}\dot{k}_x = \gamma^{-1}A\dot{\phi}(\cos\phi - \phi\sin\phi), \\
G_y &= \gamma^{-1}\dot{k}_y = \gamma^{-1}A\dot{\phi}(\sin\phi + \phi\cos\phi).
\end{aligned}
\tag{7.4.4}
$$

The equation describing the linear "velocity" along a spiral trajectory in k-space can be written as

$$
v = \gamma\sqrt{G_x^2 + G_y^2} = A\dot{\phi}\sqrt{1 + \phi^2}.
\tag{7.4.5}
$$

From Eq. (7.4.5) it follows that if $v = $ const then at large values of t

$$
\phi(t) \approx \beta\sqrt{t},
\tag{7.4.6}
$$

where β is defined by the number of revolutions and acquisition (readout) time, T_s, required to reach k_{max}: $\beta = 2\pi N_{rev}/\sqrt{T_s}$. Spirals with constant linear velocity (Figure 7.10) are used frequently because they provide high SNR [8]. Notice that Eq. (7.4.6) cannot be used near the origin, where it requires infinite gradient strength and slew rate. In practice, gradient waveforms for spiral imaging are designed to provide desirable sampling near the origin of k-space while satisfying the requirements imposed by the maximum slew rate and gradient strength available.

To estimate readout time in spiral imaging with $L \times L$ field-of-view and bandwidth ν_s, we consider constant linear velocity spirals: $v = \gamma G_0 = $ const. In this case we have $\gamma G_0 = 2\pi\nu_s/L$. If $2\pi N_{rev} \gg 1$, then we obtain the following estimate for the readout time, T_s:

$$
T_s = \frac{\pi N^2}{4M\nu_s}.
\tag{7.4.7}
$$

For example, in the case of spiral imaging with 128×128 imaging matrix, 4 interleaves, and 128 kHz bandwidth, we obtain $T_s \approx 25$ msec. In comparison, the readout time in conventional imaging with the same matrix size and bandwidth will be only 1 msec. As a result of their long readout time, spiral images display more blurring due to magnetic field nonuniformity than conventional images. Blurring in spiral images can be reduced by using image processing techniques [21, 22]. Likewise, it is reduced by increasing the number of interleaves thereby decreasing the readout time. The major limitation of the latter approach is that it prolongs scan time which is linearly dependent on the number of interleaves.

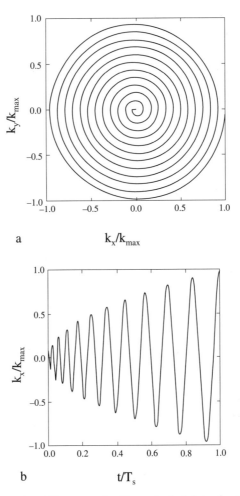

a k_x/k_{max}

b t/T_s

Figure 7.10. A constant linear velocity spiral: (a) trajectory in k-space; (b) time dependence of k_x.

Because k-space data collected on spiral trajectories do not lie on a Cartesian grid, spiral image reconstruction is more complicated than the reconstruction procedure in conventional imaging with Cartesian k-space trajectories. The acquired spiral data need to be interpolated first onto a Cartesian grid in order to subsequently implement computationally efficient Fast Fourier Transform. A commonly used algorithm for reconstructing images from k-space data sampled along spiral trajectories is described in [23].

7.5. PARTIAL K-SPACE ACQUISITION

MR imaging with partial k-space acquisition allows significant reduction in scan time as compared to standard MR imaging by acquiring only a fraction of full k-space data. Two of the frequently used partial k-space acquisition schemes, Partial Fourier (PF) [24–26] and keyhole imaging [27], are discussed in the following sections.

Partial Fourier Imaging

From the properties of Fourier transform it follows that in the case when the image intensity is real the corresponding signal, $S(k_x, k_y)$, is Hermitian. That is,

$$S(-k_x, -k_y) = S^*(k_x, k_y), \qquad (7.5.1)$$

where * denotes complex conjugation. In PF methods only a fraction of the full data set is acquired and the remainder of the data is synthesized through the use of Hermitian symmetry. PF imaging with acquisition of one half of k-space was employed by Feinberg *et al.* [24] to reduce the number of phase encodings without decreasing spatial resolution. The resulting penalty was a reduction in SNR by a factor of $\sqrt{2}$ because only half of the signals were actually acquired. If such decrease in SNR can be tolerated, this approach becomes an attractive option for rapid MRI because it reduces scan time by a factor of 2. Another application of PF methods is fractional echo acquisition, often used to shorten TE and readout time in gradient-echo imaging.

In practice various factors (e.g., field inhomogeneities in the sample, eddy currents, etc.) cause the acquired data to be non-Hermitian. Several methods have been suggested in order to reconstruct MR images from partially acquired non-Hermitian data. The following discussion focuses on one of the earliest and most effective methods proposed by Margosian [25,26] (a review of this and related methods can be found in [28]).

To describe the Margosian method we consider a one-dimensional complex image

$$I(x) = |I(x)|e^{j\phi(x)}, \qquad (7.5.2)$$

where $|I(x)|$ and $\phi(x)$ are the magnitude and phase of $I(x)$, respectively. Suppose that Fourier transform of $I(x)$

$$S(k) = \int\limits_{-\infty}^{\infty} I(x) \exp(jkx)\, dx \qquad (7.5.3)$$

is known for $k \geq 0$. Inverse Fourier transform of $S(k)$ using only half the data defines a new function, $I_{half}(x)$. It can be shown that $I_{half}(x)$ can be expressed as follows:

$$I_{half}(x) = \frac{1}{2\pi} \int\limits_{0}^{\infty} S(k) \exp(-jkx)\, dk = \frac{I(x)}{2} + \frac{j}{2\pi} \int\limits_{-\infty}^{\infty} \frac{I(x')}{x' - x} dx'. \quad (7.5.4)$$

If the image phase is known, then by multiplying $I_{half}(x)$ by $e^{-j\phi(x)}$ we obtain:

$$I_{half}(x)\, e^{-j\phi(x)} = \frac{|I(x)|}{2} + \frac{j}{2\pi} \int\limits_{-\infty}^{\infty} \frac{|I(x')|e^{j[\phi(x')-\phi(x)]}}{x' - x} dx'. \quad (7.5.5)$$

The Margosian method is based on the assumption that $\phi(x)$ is a slowly varying function such that the difference $|\phi(x) - \phi(x')|$ is small and can be neglected. In this case the magnitude image can be obtained by displaying the real part of $I_{half}(x)\, e^{-j\phi(x)}$ in (7.5.5). To calculate slowly varying phase in 2D imaging with conventional rectilinear k-space trajectories we can use a low resolution image reconstructed from a small number of central lines in k-space. In summary, the basic reconstruction algorithm in the Margosian method includes:

(a) Acquisition of about 60–75% of phase-encoded signals and zero filling of all unsampled locations in k-space
(b) Fourier transform of k-space data resulting in a complex image, $\tilde{I}(x, y)$
(c) Reconstruction of a low resolution phase map
(d) Calculation of the magnitude image, $|I(x, y)|$, by multiplying the complex image intensity, $\tilde{I}(x, y)$, in each pixel by the factor $e^{-j\phi}$ and taking the real part of the product

The Margosian method is frequently used for rapid MRI because it can easily be implemented with good results (Figure 7.11). However, this approach can sometimes cause image distortions for two main reasons. First, image phase can have significant contribution from high spatial frequency components that cannot be determined from a low resolution image. Second, the presence of large spatial variations in phase can violate the assumption that the difference $|\phi(x) - \phi(x')|$ in (7.5.5) is small.

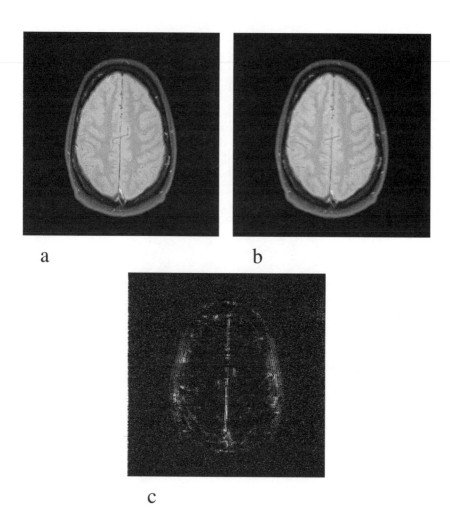

a b

c

Figure 7.11. (a) Axial 256×256 spin-echo image of the head; (b) image of the same slice reconstructed from 65% of the phase encodings by using the Margosian method; (c) difference between (a) and (b).

Dynamic Keyhole Imaging

To describe this approach it is convenient to express image intensity, I, at some arbitrary time t as a sum:

$$I(\mathbf{r}, t) = I(\mathbf{r}, t_0) + \Delta I(\mathbf{r}, t). \qquad (7.5.6)$$

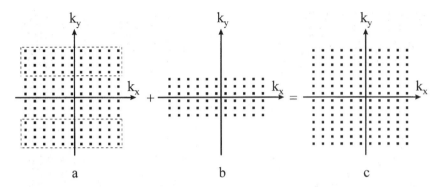

Figure 7.12. Schematic of keyhole imaging: (a) reference data; (b) dynamically acquired data; (c) resultant data set produced by combining the outlined portions of the reference data set with the dynamically acquired data.

In this equation $\mathbf{r}$ indicates spatial location in the sample, $I(\mathbf{r}, t_0)$ denotes the reference image intensity, and $\Delta I(\mathbf{r}, t)$ denotes the temporal change in intensity. The keyhole method is based on the assumption that $\Delta I(\mathbf{r}, t)$ is primarily composed of slow varying spatial components. Clearly, under this assumption $\Delta I(\mathbf{r}, t)$ can be reconstructed accurately with low spatial resolution. As a result dynamic keyhole imaging frequently acquires only low spatial frequency components of the signal. All unsampled locations in k-space eventually are filled by using the data from the reference images. In practice, keyhole images are reconstructed from dynamically collected central k-space lines corresponding to small values of the phase-encoding gradient and peripheral lines from the reference images (Figure 7.12).

Due to a reduced number of phase encodings, dynamic keyhole imaging allows significantly increased temporal resolution at the expense of reduced spatial resolution. In general, the keyhole approach is adequate if changes in intensity, $\Delta I(\mathbf{r}, t)$, vary slowly in the field-of-view. Conversely, any sharp spatial variations in intensity appearing as a result of perturbations (e.g., due to motion or administered contrast agents) of the specimen will be blurred and will cause truncation artifacts due to insufficient resolution in the phase-encoding direction (Figure 7.13).

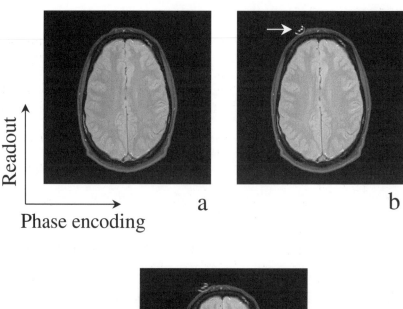

Readout →

Phase encoding →

a

b

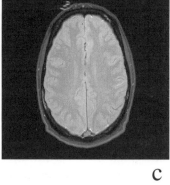

c

Figure 7.13. (a) Axial 256×256 spin-echo image of the head; (b) image acquired with the same parameters as in (a) after an intervention was simulated by inserting fluid-filled tubing (indicated by an arrow) into the field of view; (c) keyhole image reconstructed from the reference image in (a) and 64 phase encodings acquired after the intervention. Note that the tubing in the keyhole image is blurred because of insufficient spatial resolution in the phase-encoding direction.

REFERENCES

[1] A. Haase, J. Frahm, D. Matthaei, W. Hanicke, K.D. Merboldt. "FLASH imaging. Rapid NMR imaging using low flip-angle pulses," *J. Magn. Reson.*, **67**, 258 (1986).

[2] P. van der Meulen, J.P. Groen, J.J. Cuppen. "Very fast MR imaging by field echoes and small angle excitation," *Magn. Reson. Imaging*, **3**, 297 (1985).

[3] A. Haase. "Snapshot Flash MRI. Applications to T_1, T_2 and chemical shift imaging," *Magn. Reson. Med.*, **13**, 77 (1990).

[4] P. Mansfield. "Multi-planar image formation using NMR spin echoes," *J. Phys., C* **10**, L55 (1977).

[5] P. Mansfield, I.L. Pykett. "Biological and medical imaging by NMR," *J. Magn. Reson.*, **29**, 355 (1978).

[6] J. Henning, A. Naureth, H. Friedburg. "Rare imaging: a fast imaging method for clinical MR," *Magn. Reson. Med.*, **3**, 823 (1986).

[7] C.B. Ahn, J.H. Kim, Z.H. Cho. "High-speed spiral-scan echo planar NMR imaging," *IEEE Trans. Med. Imag.*, **1**, 2 (1986).

[8] G.H. Meyer, B.S. Hu, D.G. Nishimura, A. Macovski. "Fast spiral coronary artery imaging, *Magn. Reson. Med.*, **28**, 202 (1992).

[9] R.R. Ernst, W.A. Anderson. "Application of Fourier transform spectroscopy to magnetic resonance," *Rev. Sci. Instr.*, 37, 93 (1965).

[10] R. Freeman, H.D.W. Hill. "Phase and intensity anomalies in Fourier transform NMR," *J. Magn. Reson.*, **4**, 366 (1971).

[11] Y. Zur, M.L. Wood, L.J. Neuringer. "Spoiling of transverse magnetization in steady-state sequences," *Magn. Reson. Med.*, **21**, 251 (1991).

[12] A. P. Crawley, M.L. Wood, R.M. Henkelman. "Elimination of transverse coherences in FLASH MRI," *Magn. Reson. Med.*, **8**, 248 (1988).

[13] J. Coremans, M. Spanogne, L. Budinsky, J. Sterckx, R. Luypaert, H. Eisendrath, M. Osteaux. "A comparison between different imaging strategies for diffusion mesurements with the centric phase-encoded TurboFLASH sequence," *J. Magn. Reson.*, **124**, 323 (1997).

[14] A. Oppelt, R. Graumann, H. Barfub, H. Fisher, W. Hartl, W. Schajor. "FISP – a new fast MRI sequence," *Electromedica*, **54**, 15 (1986).

[15] K. Sekihara. "Steady-state magnetization in rapid NMR imaging using small flip angles and short repetition intervals," *IEEE Trans. Med. Imaging*, **MI–6**, 157 (1987).

[16] W.S. Hinshow. "Image formation by nuclear magnetic resonance: the sensitive-point method," *J. Appl. Phys.*, **47**, 3709 (1976).

[17] E.O. Stejskal, J.E. Tanner. "Spin diffusion measurements: spin echoes in the presence of a time-dependent field gradient," *J. Chem. Phys.*, **42**, 288 (1965).

[18] P.S. Melki, F.A. Jolesz, R.V. Mulkern. "Partial RF echo planar imaging with the FAISE method. I. Experimental and theoretical assessment of artifact," *Magn. Reson. Med.*, **26**, 328 (1992).

[19] R.V. Mulkern, P.S. Melki, P. Jakab, N. Higuchi, F.A. Jolesz. "Phase-encode order and its effect on contrast and artifact in single-shot RARE sequences," *Med. Phys.*, **18**, 1032 (1991).

[20] K.K. King, T.K.F. Foo, C.R. Crawford. "Optimized gradient waveforms for spiral scanning," *Magn. Reson. Med.*, **34**, 156 (1995).

[21] D.C. Noll, C.H. Meyer, J.M. Pauly, D.G. Nishimura, A. Macovski. "A homogeneity correction method for magnetic resonance imaging with time-varying gradients," *IEEE Trans. Med. Imaging*, **10**, 629 (1991).

[22] P. Irarrazabal, C.H. Meyer, D.G. Nishimura, A. Macovski. "Inhomogeneity correction using an estimated linear field map," *Magn. Reson. Med.*, **35**, 278 (1996).

[23] J. Jackson, C. Meyer, D. Nishimura, A. Macovski. "Selection of a convolution function for Fourier inversion using gridding," *IEEE Trans. Med. Imaging*, **10**, 473 (1991).

[24] D.A. Feinberg, J.D. Hale, J.C. Watts, L. Kaufman, A. Mark. "Halving MR imaging time by conjugation: demonstration at 3.5 kG," *Radiology*, **161**, 527 (1986).

[25] P. Margosian. "Faster MR imaging—imaging with half the data," *Proceeding SMRM*, 1024 (1985).

[26] P. Margosian, F.Schmitt, D.E. Purdy. "Faster MR imaging: imaging with half the data," *Health Care Instr.*, **1**, 195 (1986).

[27] J.J. van Vaals, M.E. Brummer, W.T. Dixon, H.H. Tuithof, H. Engels, R.C. Nelson, B.M. Gerety, J.L. Chezmar, J.A. Boer. " 'Keyhole' method for accelerating imaging of contrast agent uptake," *JMRI*, **3**, 671 (1993).

[28] Z.-P. Liang, F.E. Boada, R.T. Constable, E.M. Haacke, P.L. Lauterbur, M.R. Smith. "Constrained reconstruction methods in MR imaging," *Rev. Magn. Reson. Med.*, **4**, 67 (1992).

CHAPTER 8

MR Imaging of Flow

The effect of flow on the observed NMR signal was first described by Suryan [1] soon after the discovery of NMR. Suryan observed an increased NMR signal from a flowing liquid as compared to the signal from the same liquid at rest. Suryan correctly explained the observed effect by the decreased saturation of the magnetization in the flowing liquid. A decade later the same effect was reported by Singer [2, 3] who, along with Bowman and Kudracev [4], developed several techniques for quantitative NMR flow measurements. The studies of these and other investigators laid the groundwork for MR flow imaging. Reference [5] contains an interesting review of early NMR studies of flow.

After flow-related effects in MR images were first discussed by Young et al. [6] and Crooks et al. [7] in the early 1980s, various techniques have been developed for MRI of flow. In general, these techniques can be divided into three major categories based on the effects used for flow imaging: (a) "time-of-flight" effects [8–21]; (b) velocity-induced phase [22, 23]; and (c) signal enhancement caused by MR contrast agents [24–30].

8.1. TIME-OF-FLIGHT TECHNIQUES

Gradient-Echo Imaging

Spatially selective r.f. pulses are used in 2D and 3D MR imaging to repeatedly excite magnetization in a slice of material. When flowing spins enter an excited slice during imaging, they normally have a higher longitudinal magnetization than that of the saturated static material in the slice. Therefore, in-flow of unsaturated spins makes the flowing material in images appear brighter than the static material. Time-of-flight (TOF) MR imaging utilizes the difference in magnetization of stationary and flowing spins in order to improve visualization of flow in MR images. Gradient-echo TOF MRI based on the in-flow effect is performed with short repetition time $TR \ll T_1$, needed for saturation of the static material in the excited slice. Flow-induced dephasing of spins is reduced by imaging with short TE (a few milliseconds) combined with fractional-echo acquisition or by employing gradient moment nulling techniques (see Chapter 6).

In 2D TOF imaging the signals are acquired from consecutively excited thin slices (1.5–3 mm thickness), which are typically chosen perpendicular to the direction of flow in order to minimize saturation of the flowing material (Figure 8.1(a)). Because it is often difficult to assess flow conditions from conventional two-dimensional MR images, additional images of flow can be produced by using a post-processing technique, known as the *maximum intensity projection* (MIP). In this technique an imaged volume is projected onto a plane. The determined maximum voxel intensities along rays perpendicular to the projection plane are displayed subsequently as two-dimensional MIP images. The MIP algorithm is very effective for selectively displaying high-intensity regions in the imaged volume. MIP images of the carotid and vertebral arteries in the neck reconstructed from images of thin slices are shown in Figure 8.1(b) and (c).

Since 3D imaging in general provides higher SNR and allows better resolution than 2D imaging, it is often chosen as the technique for diagnostic MR imaging of arteries, known as *magnetic resonance angiography* (MRA). Because in 3D imaging signals are acquired following excitation of a relatively thick slice (slab), the flowing material (blood) is continuously saturated by successive excitation pulses as it propagates through the slab. This saturation effect is particularly significant under the conditions of slow flow or flow travelling parallel to the imaged slab. The saturation of blood is spatially dependent, varying from its minimum near the entrance to the slab to its maximum near the exit (Figure 8.2). Depending on the flow velocity, TR, and flip angle, this effect can lead to noticeable variations in the intensity of blood in 3D TOF images. Several imaging techniques have recently been suggested in order to ameliorate contrast in 3D TOF MRA. One of these techniques employs r.f. excitation with a spatially varying flip angle that increases from its minimum value at the slab entrance to

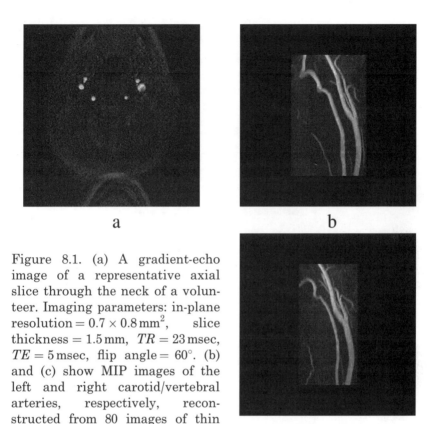

a

b

Figure 8.1. (a) A gradient-echo image of a representative axial slice through the neck of a volunteer. Imaging parameters: in-plane resolution $= 0.7 \times 0.8\,\mathrm{mm}^2$, slice thickness $= 1.5\,\mathrm{mm}$, $TR = 23\,\mathrm{msec}$, $TE = 5\,\mathrm{msec}$, flip angle $= 60°$. (b) and (c) show MIP images of the left and right carotid/vertebral arteries, respectively, reconstructed from 80 images of thin slices (like the one shown in (a)).

c

its maximum value at the slab exit [13]. Example of MIP images of the brain reconstructed from 3D TOF data acquired with varying flip angle are shown in Figure 8.3. In another approach saturation of flowing material in 3D TOF MRA is decreased by sequential imaging of overlapping thin slabs [14].

In many instances the close proximity of veins to arteries might make it difficult to differentiate between the two in MR images. Visualization of arterial structures in TOF MRA can be improved by selectively saturating the magnetization of venous blood (see Chapter 6). The use of selective saturation in TOF MRA is facilitated by the fact that the direction of arterial blood flow is typically opposite to that of venous blood flow. In practice the magnetization of venous blood is reduced routinely by selectively saturating a slab of tissue through which the blood flows prior to entering the imaged volume

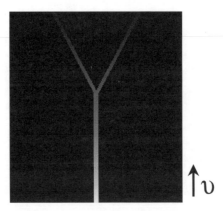

Figure 8.2. Simulated image demonstrating spatial variations in intensity of flow in 3D TOF imaging caused by saturation of the flowing material in the imaged slab. The saturation of the flow continuously increases as it propagates through the slab.

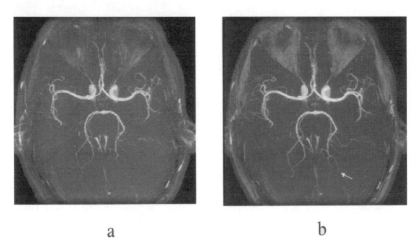

a   b

Figure 8.3. (a) MIP image of cerebral vessels in the brain reconstructed from a data set obtained with 3D acquisition. Imaging parameters: spatial resolution $= 0.6 \times 0.8 \times 0.8 \, \text{mm}^3$, $TR = 40 \, \text{msec}$, $TE = 3 \, \text{msec}$, average flip angle $= 30°$. (b) MIP image reconstructed from a data set acquired with the same parameters as in (a). In addition, the pulse-sequence used in (b) included an off-resonance preparation pulse providing saturation of the macromolecular protons (see Chapter 4). The resulting MTC enhances visualization of small vessels (like the one indicated by an arrow in (b)).

(Figure 6.10). The resulting effect is the dark appearance of veins relative to the bright appearance of arteries in TOF images.

Spin-Eecho Imaging

Initially excited blood remains in the imaging volume during the time $\tau = d/v$, where d is the slice thickness and v is the component of the flow velocity perpendicular to the slice. If the time interval $TE/2$ between excitation and the spatially selective 180 degree pulse in spin-echo imaging exceeds τ, then only the magnetization of the static material is refocused. Therefore, the washout of flowing spins from the imaged slice, known as the *out-flow effect*, makes blood vessels appear darker than the surrounding tissue in the image. For example, spin-echo images acquired with slice thickness of 2 mm, TE of 20 ms would display a greatly reduced signal from flow (flow void) if the flow velocity exceeds 20 cm/sec. Imaging techniques based on the out-flow effect [15] are frequently referred to as "black blood" MRA because of the darker appearance of blood relative to that of stationary material in the MRA images. Conversely, techniques used to acquire images with bright appearance of blood are known as "bright blood" MRA. Unlike bright blood MRA, black blood MRA techniques generate flow contrast by minimizing the signal from moving spins. As a result, these techniques also minimize artifacts due to blood flow pulsations, saturating of the blood magnetization and flow-induced dephasing of spins which are often present in bright blood MRA images.

Inversion-Recovery Imaging

To improve contrast between flowing and stationary spins, Nishimura *et al.* [16] suggested a subtraction technique that can completely eliminate signal from the static material. In this technique two images are acquired as follows: the first image is obtained following a 180 degree pulse that selectively inverts magnetization of the spins in the slab and above; the second image is acquired after the magnetization in the entire volume is inverted (Figure 8.4). In both cases the same time delay, TI, is used between the inversion and excitation pulses. Image subtraction leads to cancellation of the signals from both moving and stationary spins that have experienced inversion pulses. The resulting image is characterized by a bright appearance of vessels with upstream flow entering the slab from below. The major advantage of this approach as compared to the gradient-echo

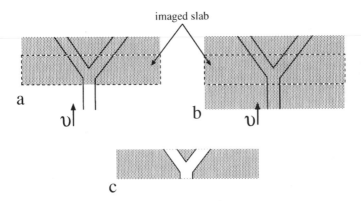

Figure 8.4. A diagram describing selective-inversion MRA with image subtraction: (a) first image is acquired following inversion of the magnetization in the slab and above; (b) second image is acquired after magnetization in the entire volume is inverted; (c) image subtraction results in cancellation of the signal from the static material in the imaged slab.

techniques described earlier is potentially better visualization of vessels with slow flow, provided that the inversion time *TI* is long enough so that complete replenishment of the flowing material in the slab takes place. The major limitations are long scan time, because two images are acquired, and increased sensitivity to patient motion during data acquisition, because image subtraction is used.

In another approach selective inversion of the longitudinal magnetization is used in order to null the signal from blood [17]. In this black blood MRA approach a nonselective 180 degree pulse first inverts the longitudinal magnetization in the entire volume. Immediately after the first inversion a spatially-selective 180 degree pulse returns the magnetization to its equilibrium state in the imaged slice only (Figure 8.5). During the inversion time, *TI*, the blood that initially experienced selective 180 degree pulse leaves the slice. The longitudinal magnetization of the blood entering the slice at time *TI* is proportional to $1 - 2e^{-TI/T_{1,blood}}$, where $T_{1,blood}$ is the spin-lattice relaxation time of blood. The flow signal is nulled by choosing the appropriate inversion time: $TI = T_{1,blood} \ln 2$.

Flow Imaging with Spatial Modulation of Magnetization

Visualization of flow-related displacements of spins in MR images can be achieved by using specially prepared magnetization [18, 19]. In

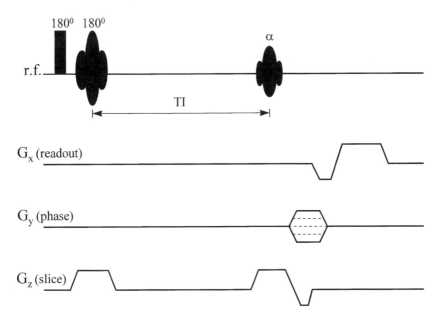

Figure 8.5. Pulse-sequence for selective-inversion black blood MR imaging.

this approach, known as *spin tagging,* the longitudinal magnetization, M_z, in the sample is modulated initially by an arbitrary periodical function. Subsequent MR imaging depicts changes in the modulation pattern caused by flow or motion.

Immediately after spin tagging the longitudinal magnetization can be expressed as

$$M_z(\mathbf{r}, t = 0) = M_0 T_{ag}(\mathbf{r}), \qquad (8.1.1)$$

where M_0 is the steady-state longitudinal magnetization and $T_{ag}(\mathbf{r})$ is an arbitrary periodical function varying between -1 and 1. Displacements of spins due to flow with velocity $\mathbf{v}$ shift the initial tagging pattern by $\mathbf{v}t$. The resultant effect of flow and T_1 relaxation on the longitudinal magnetization is described by the following equation:

$$M_z(\mathbf{r}, t) = M_0 e^{-t/T_1}[T_{ag}(\mathbf{r} - \mathbf{v}t) - 1] + M_0. \qquad (8.1.2)$$

Periodical modulation of M_z can be produced by using several spin-tagging techniques [18–20]. A basic spin-tagging pulse sequence includes a pair of nonselective 90 degree pulses (Figure 8.6). After the first 90 degree pulse the phase of the transverse magnetization, ϕ,

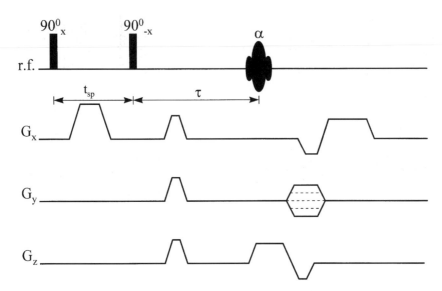

Figure 8.6. Basic pulse sequence for spin-tagging imaging of flow.

changes under the influence of the applied gradient, G_x: $\phi = \gamma x \int G_x \, dt$. Under the assumption that the relaxation and flow effects during spin-tagging can be neglected, the longitudinal magnetization immediately after the second 90 degree pulse is given by

$$M_z = M_0 \cos kx, \qquad (8.1.3)$$

where $k = \gamma \int_0^{t_{sp}} G_x \, dt$. Any residual transverse magnetization is subsequently destroyed by crusher gradients applied prior to imaging. If two-dimensional spatial modulation of the magnetization is needed, it can be produced by applying a pair of orthogonal gradients (e.g., G_x and G_y) during the time interval between the 90 degree pulses.

Following spin tagging the modulated nuclear magnetization is imaged. By using equations (8.1.2) and (8.1.3) we obtain that the transverse magnetization in the sample during imaging can be written as

$$M_{tr} \propto M_0 e^{-\tau/T_1} \cos k(x - v_x \tau) + M_0 (1 - e^{-\tau/T_1}), \qquad (8.1.4)$$

where v_x is the x-component of the flow velocity and τ is the interval between the spin-tagging and imaging parts of the pulse sequence. The spatially varying transverse magnetization appears as a series of bright bands in images (Figure 8.7(a)). In principle, local displacements of spins during the interval τ can be assessed by examining distortions (bending) of the initially straight bands (Figure 8.7(b)).

a   b

Figure 8.7. (a) Axial 256×256 spin-echo image of a glass container filled with poppy seeds. The image was acquired after a delay of 120 msec following modulation of the magnetization in the container. The distance between bright bands in the image is approximately 3 mm. (b) Image of the container acquired with the same parameters as in (a) but after the flow of seeds in the container was induced by periodical shaking. Reprinted with permission from [21].

8.2. PHASE-CONTRAST MRA

Phase-contrast imaging of flow utilizes the phase of magnetization in order to differentiate between moving and static materials [22, 23]. In this approach phase encoding of flow is implemented by using bipolar magnetic field gradients. A bipolar gradient, G_x, can be defined as

$$G_x = \begin{cases} -G_v, & t_0 \leq t \leq t_0 + \tau \\ G_v, & t_0 + T \leq t \leq t_0 + T + \tau \\ 0, & \text{elsewhere} \end{cases} \tag{8.2.1}$$

where G_v, τ and T are the amplitude, duration, and separation of the two gradient lobes, respectively (Figure 8.8).

Assuming that the flow velocity, v_x, in the direction of the applied bipolar gradient is constant, the accumulated phase shift can be written as

$$\phi = \int_{t_0}^{t_0 + T + \tau} \gamma G_x(t) x(t) \, dt = \gamma v_x G_v \tau T. \tag{8.2.2}$$

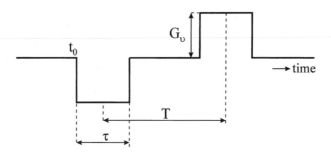

Figure 8.8. A bipolar gradient used for phase-contrast imaging of flow.

This equation shows that the flow-induced phase is directly proportional to velocity. Consequently, the overall phase shift for stationary material $(v_x = 0)$ is zero.

In phase-contrast MRA, bipolar gradients are incorporated into MRI pulse sequences. The velocity-dependent phase can be used for flow imaging if the pulse sequence is repeated with the inverted bipolar gradient (Figure 8.9) and the signal from static material is cancelled out by calculating the difference between the reconstructed images:

$$\Delta I = I_v e^{j\phi} - I_v e^{-j\phi} = 2j I_v \sin \phi. \qquad (8.2.3)$$

In this equation I_v is a proportionality factor, which is linearly dependent on the number of moving spins in a voxel. MR images of flow can

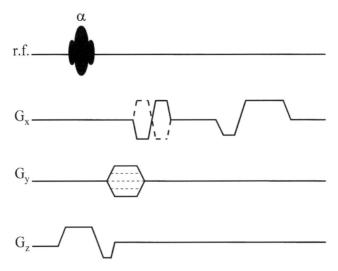

Figure 8.9. Gradient-echo pulse sequence for phase-contrast MRA.

be produced by displaying the magnitude of ΔI. Because of the non-linear relationship between ΔI and ϕ, bipolar gradients need to be experimentally adjusted to insure that $\phi < 1$. Under this condition $\sin \phi \approx \phi$ and ΔI is approximately proportional to the component of flow in the direction of the applied bipolar gradient.

In many instances it might be desirable to obtain images in which local image intensity is proportional to the magnitude of flow rather than components of flow. A measure of the flow magnitude can be obtained if the reconstructed images of orthogonal x, y and z-components of flow (i.e., ΔI_x, ΔI_y and ΔI_z, respectively) are combined as follows: $\Delta I = \sqrt{|\Delta I_x|^2 + |\Delta I_y|^2 + |\Delta I_z|^2}$. The limitation of this approach is a significant increase in scan time because six images must be acquired before an image of the flow magnitude can be produced.

The phase-contrast technique for flow imaging described above is well suited to imaging uniform flow. However, it cannot be used to measure the velocity distribution function of spins. This limitation can be overcome by using a general method for Fourier velocity encoding described by Moran [22]. In this method the strength of the applied bipolar gradient is incremented N times in a step-like fashion: $G_v = nG_0$, $-N/2 \leq n < N/2$. Consequently, the velocity-dependent phase of the magnetization becomes dependent on n. For example, when imaging the x-component of flow the velocity-dependent phase is given by

$$\phi(n) = \gamma G_v \tau T v_x = \frac{\pi n v_x}{V_g}, \qquad (8.2.4)$$

where $V_g = \pi/\gamma G_0 \tau T$. The NMR signal generated by spins with velocity distribution function $\psi(v_x)$ can then be written as

$$S(n) \propto \int \psi(v_x) e^{j \pi n v_x / V_g} \, dv_x. \qquad (8.2.5)$$

According to this equation the signal is proportional to the Fourier transform of $\psi(v_x)$. Under the assumption that V_g exceeds the maximum velocity of spins, the velocity distribution function can be reconstructed with resolution of $2V_g/N$ by computing inverse Fourier transform of $S(n)$. It should be noted that this velocity encoding procedure results in a substantial increase in scan time (i.e., by a factor of N) when used in combination with techniques for spatial encoding. This practical limitation often makes it difficult to utilize the vast potential of the Fourier velocity encoding for MR imaging *in vivo*.

8.3. CONTRAST-ENHANCED MRI OF FLOW

Recently, a number of studies have demonstrated that contrast between static tissue and blood in MR angiography can be improved by using gadolinium-containing contrast agents that significantly reduce the spin-lattice relaxation time of blood [24–27]. The shortening of T_1 results in decreased saturation of the blood magnetization, leading to improved visualization of blood vessels in MR images. Because this technique, also known as *contrast-enhanced MRA*, relies on the effect of T_1 shortening it is less flow dependent than TOF MRA. As a result, contrast-enhanced MRA mostly avoids artifacts due to saturation of blood magnetization, which often degrade bright blood MRA images.

Currently, contrast-enhanced MRI of flow is used primarily for imaging of the arterial system (Figure 8.10). Practical implementation of contrast-enhanced MRA is based on the use of fast 3D gradient-echo imaging with the shortest possible *TR*. Short scan time in contrast-enhanced MRA is needed to: (a) acquire signals during the so-called arterial phase (~10 sec) when the contrast media exists primarily in the imaged arteries but not in the adjacent veins; (b) minimize the effect of contrast media leakage from the blood into the surrounding tissue; (c) enable breathholding which tends to minimize image distortion due to respiratory motion. Because the arterial phase is usually

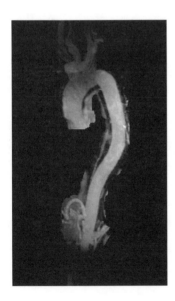

Figure 8.10. MIP image of the aorta produced from a 3D data set acquired during 34 sec after administration of Gd-DTPA. Imaging parameters: spatial resolution = $1.5 \times 1.75 \times 1.3 \, \text{mm}^3$, $TR = 7 \, \text{msec}$, $TE = 2 \, \text{msec}$, flip angle = $45°$.

shorter than the typical scan time in 3D MRA, it is a common practice to acquire only signal components with low spatial frequencies (i.e., central portion of k-space) during the arterial phase while components with high spatial frequencies are acquired after contrast media has started to accumulate in the veins and soft tissues. Acquiring the central portion of k-space too early (i.e., before the arrival of the bolus of contrast media) or too late (i.e., after the subsequent enhancement of veins and soft tissue) has the adverse effect of decreased contrast between the arteries and their surroundings. To achieve the optimum image contrast, acquisition of components with low spatial frequencies should start immediately after the bolus of contrast media has arrived in the imaged arterial structures. Two approaches have recently been suggested in order to determine the arrival time of administered contrast material [28–30]. In the first approach, the time course of the vascular signal is monitored after administration of a small test bolus of contrast media. The determined arrival time is then used as a time delay between the subsequent administration of the full bolus and the beginning of the signal acquisition [28]. An alternative approach is based on monitoring the signal from a tracking volume that can include or be adjacent to the imaged site [29, 30]. The MRA signal acquisition can be triggered either manually when contrast arrival is detected in the images of the tracking volume [29] or automatically when the signal from the tracking volume exceeds a chosen threshold [30].

REFERENCES

[1] G. Suryan. "Nuclear resonance in flowing liquids," *Proc. Indian. Acad. Sci.*, Section A. **33**, 107 (1951).

[2] J.R. Singer. "Blood flow rates by nuclear magnetic resonance measurements," *Science*, **130**, 1652 (1959).

[3] J.R. Singer. "Flow rates using nuclear or electron paramagnetic resonance techniques with applications to biological and chemical processes," *J. Appl. Phys.*, **31**, 125 (1960).

[4] R.L. Bowman, V. Kudravcev. "Blood flowmeter utilizing NMR," *IRE Trans. Med. Electron.*, **ME-6**, 267 (1959).

[5] J.H. Battocletti. "Blood flow measurements by NMR," *CRC Critic. Rev. Biom. Engin.*, **13**, 311 (1986).

[6] I.R. Young, M. Burl, G.J. Clarke, A.S. Hall, T. Pasmore, A.G. Collins, D.T. Smith, J.S. Orr, G.M. Bydder, F.H. Doyle, R.H. Greenspan, R.E. Steiner. "Magnetic resonance properties of hydrogen: imaging the posterior fossa," *AJR*, **137**, 895 (1981).

[7] L. Crooks, P. Sheldon, L. Kaufman, W. Rowan, T. Miller. "Quantification of obstructions in vessels by Nuclear Magnetic Resonance (NMR)," *IEEE Trans. Nucl. Science*, **NS-29**, 1181 (1982).

[8] G.T. Gullberg, F.W. Wehrli, A. Shimakawa, M.A. Simons. "MR vascular imaging with a fast gradient refocusing pulse sequence and reformatted images from transaxial sections," *Radiology*, **165**, 241 (1987).

[9] P.J. Keller, B.P. Drayer, E.K. Fram, K.D. Wiliams, C.L. Dumoulin, S.P. Souza. "MR angiography with two-dimensional acquisition and three-dimensional display," *Radiology*, **173**, 527 (1989).

[10] G.W. Laub, W.A. Kaiser. "MR angiography with gradient motion refocusing," *J. Comput. Assisst. Tomogr.*, **12**, 377 (1988).

[11] C.L. Dumoulin, H.E. Cline, S.P. Souza, W.A. Wagle, M.F. Walker. "Three-dimensional time-of-flight magnetic resonance angiography using spin saturation," *Magn. Res. Med.*, **11**, 35 (1989).

[12] P.M. Ruggieri, G.A. Laub, T.J. Masaruk, M.T. Modic. "Intracranial circulation: pulse-sequence consideration in three-dimensional (volume) MR angiography," *Radiology*, **171**, 785 (1989).

[13] T. Nagele, U. Klose, W. Grodd, W. Peterson, J. Tintera. "The effects of linearly increasing flip angles on 3D inflow MR angiography," *Magn. Reson. Med.*, **31**, 561 (1994).

[14] D.L. Parker, C. Yuan, D.D. Blatter. "MR angiography by multiple thin slab 3D acquisition," *Magn. Reson. Med.*, **17**, 434 (1991).

[15] R.R. Edelman, H.P. Mattle, B. Wallner, R. Bajakian, J. Kleefield, C. Kent, J. Skillman, J.B. Mendell, D.J. Atkinson. "Extracranial carotid arteries: evaluation with "black blood" MR angiography," *Radiology*, **177**, 45 (1990).

[16] D.G. Nishimura, A. Macovski, J.M. Pauly, S.M. Conolly. "MR angiography by selective inversion recovery," *Magn. Reson. Med.*, **4**, 193 (1987).

[17] R.R. Edelman, D. Chien, D. Kim. "Fast selective black blood MR imaging," *Radiology*, **181**, 655 (1991).

[18] E.A. Zerhouni, D.M. Parish, W.J. Rogers, A. Yang, E.P. Shapiro. "Human heart: tagging with MR imaging – a method for noninvasive assessment of myocardial motion," *Radiology*, **169**, 59 (1988).

[19] L. Axel, L. Dougherty. "MR imaging of motion with spatial modulation of magnetization," *Radiology*, **171**, 841 (1989).

[20] T.J. Mosher, M.B. Smith. "A DANTE tagging sequence for the evaluation of the translational sample motion," *Magn. Reson. Med.*, **15**, 334 (1990).

[21] V.Yu. Kuperman, E.E. Ehrichs, H.M. Jaeger, G.S. Karczmar. "A new technique for differentiating between diffusion and flow in granular media using magnetic resonance imaging," *Rev. Sci. Instrum.*, **66(8)**, 4350 (1995).

[22] P.R. Moran. "A flow velocity zeugmatographic interlace for NMR imaging in humans," *Magn. Reson. Imag.*, **1**, 197 (1982).

[23] C.L. Dumoulin, H.R. Hart. "Magnetic resonance angiography," *Radiology*, **161**, 717 (1986).

[24] M.R. Prince, E.K. Yucel, J.A.Kaufman, D.C. Harrison, S.C. Geller. "Dynamic gadolinium-enhanced 3DFT abdominal MR arteriography," *J. Magn. Reson. Imag.*, **3**, 877 (1993).

[25] M.R. Prince. "Gadolinium-enhanced MR aortography," *Radiology*, **191**, 155 (1994).

[26] M.K. Adams, W. Li, P.A. Wielopolski, D. Kim, E.J. Sax, K.C. Kent, E.R. Edelman. "Dynamic contrast-enhanced subtraction MR angiography of the lower extremities: initial evaluation with multisection two-dimensional time-of-flight sequence," *Radiology*, **196**, 689 (1995).

[27] M.T. Alley, R.Y. Shifrin, N.J. Pelc, R.J. Herfkens. "Ultrafast contrast-enhanced three-dimensional MR angiography: state of the art," *RadioGraphics*, **18**, 273 (1998).

[28] J.P. Earl, N.M. Rofsky, D.R. DeCorato, G.A. Krinsky, J.C. Weinreb. "Breath-hold single-dose gadolinium-enhanced three-dimensional MR aortography: usefulness of a timing examination and MR power injector," *Radiology*, **201**, 705 (1996).

[29] A.H. Wilman, S.J. Riederer, B.F. King, J.P. Debbins, P.J. Rossman, R.L. Ehman. "Fluoroscopically-triggered contrast-enhanced 3D MR angiography with elliptical centric view order: application to the renal arteries," *Radiology*, **205**, 137 (1997).

[30] T.K.F. Foo, M. Saranathan, M.R. Prince, T.L. Chenevert. "Automated detection of bolus arrival and initiation of data acquisition in fast, three-dimensional, gadolinium-enhanced MR angiography," *Radiology*, **203**, 275 (1997).

CHAPTER 9

MRI Instrumentation: Magnets, Gradient Coils, and Radiofrequency Coils

MR imaging requires three different types of magnetic fields. First, a static and highly uniform magnetic field, known as the *main field*, is required to create the initial longitudinal magnetization in the object and to maintain Larmor precession of nuclear spins at constant angular frequency. Second, an r.f. field is required for excitation of the transverse magnetization in the object. Finally, highly linear magnetic field gradients that can be turned on and off during a relatively short time (normally less than 1 msec) are essential for spatial localization in MR imaging. Because of the different functions and required characteristics of the main field, r.f. field, and gradient field, they are produced by different devices known as magnets, r.f. coils, and gradient coils, respectively. The purpose of this chapter is to describe main types and characteristics of these devices.

9.1. MAGNETS

The purpose of magnets used for MRI is to produce a static, strong, and homogeneous magnetic (main) field, $\mathbf{B_0}$. Before discussing the different types of magnets later in this section, it is instructive to first consider the required characteristics of the main field used for NMR imaging and spectroscopy.

Main Field Stability and Strength

During MR imaging or spectroscopic studies, signals are acquired repeatedly over time following successive excitations of the transverse magnetization in a sample. As a result, temporal stability of the main magnetic field is extremely important. The cause of temporal variations in B_0 can vary depending upon the magnet type and design. For example, main field in a resistive magnet can fluctuate because of the fluctuations in the supplied power. Consequently, power supplies for resistive magnets must be extremely stable. Main field in a permanent magnet is extremely sensitive to temperature fluctuations, which makes it necessary to monitor and control the temperature of the magnet. Compared with permanent and resistive magnets, superconducting magnets provide the greatest temporal stability in operation with temporal variations in B_0 less than $B_0 \times 10^{-7}$ per hour.

The majority of clinical whole-body MRI scanners operate at field strengths between 0.3 Tesla and 2 Tesla. More recently, the need to increase the sensitivity of functional MRI of the brain has stimulated interest in proton imaging at field strengths in the 3–8 Tesla range. High field strengths offer the advantage of increased signal-to-noise ratio in both NMR imaging and spectroscopy. In addition for spectroscopic applications, higher field strengths make it easier to resolve individual lines in NMR spectra because the difference between Larmor frequencies of the same nuclei in different chemical environments (chemical shift) increases linearly with field strength.

Main Field Homogeneity

Magnetic field inhomogeneity is detrimental for the quality of MR images because of the associated geometric distortion and signal loss. Field inhomogeneity is particularly troublesome for NMR spectroscopy because it might not allow differentiation between closely located individual lines in the NMR spectra. The degree of magnetic field homogeneity is commonly defined in terms of the normalized deviation of the field in a given spherical volume:

$$\frac{\delta B}{B_c} = \frac{B_{max} - B_{min}}{B_c}. \tag{9.1.1}$$

In this equation B_{min} and B_{max} are the minimum and maximum field strengths in the spherical volume, respectively, and B_c is the field strength at the center of the sphere. Magnetic field homogeneity is normally expressed in parts per million (ppm), $\delta B/B_c \times 10^6$. In practice, main field homogeneity for different types of magnets is typically on the order of 10 ppm for a 50 cm diameter spherical volume (dsv). Compared to permanent and resistive magnets, superconducting magnets offer the highest field homogeneity (about 5 ppm in a 50 cm dsv for commercially available systems). The required homogeneity of the main field can vary depending on the size of the specimen as well as the type and objectives of NMR studies. For example, *in vivo* NMR spectroscopy may require homogeneity better than 0.1 ppm for a 10 cm dsv [1].

Permanent Magnets

The unique property of permanent magnet materials is the existence of microscopic regions, known as *magnetic domains*, with nonzero magnetic moments. In their natural state the magnetic moments of different domains are randomly oriented; however, in the presence of an external magnetic field, the domain moments become aligned along the field. It is remarkable that the alignment of magnetic moments remains after the magnetizing field is removed, making it possible to create permanent magnets from these materials.

Modern materials for permanent magnets include neodymium–iron–boron alloys (NdFeB), ceramic ferrite, alnico (aluminum–nickel–iron–cobalt) alloys and samarium–cobalt alloys. The performance of a given magnetic material depends on two main parameters: the strength of an external magnetic field required to demagnetize the material, known as the *coercivity*, and the maximum achievable product of magnetic field and magnetic induction, known as *the energy product*. NdFeB and SmCo materials offer the highest energy product and coercivity among different permanent magnet materials currently used.

Permanent magnets for whole-body MR scanners have several important limitations. First, the maximum field strength currently feasible for permanent magnets is relatively low (0.2–0.3 Tesla). Second, permanent magnets are very sensitive to temperature fluctuations. For example, the temperature coefficient of magnetic field for NdFeB is $-0.12\%/°C$ [2]; therefore, a change in temperature by only $0.1°C$ can cause a 120 ppm change in magnetic field. Because of such high temperature sensitivity, practical use of permanent magnets requires a special system for temperature monitoring and regulation. Other limitations of permanent magnets include their heavy weight (which increases with field strength)[1] and high cost of permanent magnet materials.

---

[1] For example, a 0.3 Tesla permanent magnet described in [3] has a mass of 90,000 kg.

Iron frame

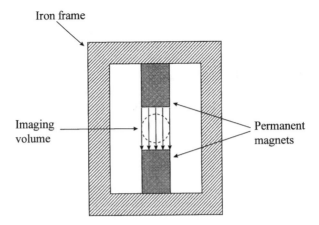

Figure 9.1. Schematic of an H-frame magnet.

A popular design for whole-body permanent magnets is known as the *H-frame*. An H-frame magnet contains an iron frame that provides a return path for magnetic flux and supports a pair of permanent magnets (poles) generating a magnetic field in the gap where the patient is positioned (Figure 9.1). The main advantages of this design are easy patient access (compared to commonly used solenoidal electromagnets) and small fringe field due to intrinsic shielding provided by the frame.

Resistive Magnets

Magnetic field in resistive magnets is produced by electric current in specially designed coils. The coils are made of aluminum or copper, materials with high electric conductivity. Two basic types of resistive magnets used for whole-body imaging are air-core magnets and iron-core magnets. The latter type employs ferromagnetic materials to increase the strength of B_0.

Magnetic field strength in an air-core resistive magnet is proportional to the operating current. Typical operating current and electrical resistance for a 0.15 Tesla air-core magnet are about 200 A and 1 ohm, respectively [3]. When in use, the magnet would require a 40 kilowatt power supply. As a result of the high power dissipating in a resistive magnet, it is necessary to use a special cooling system that absorbs the produced heat and ensures the required stability of the magnet temperature. The latter is extremely important because temperature fluctuations can alter the conductivity of the magnet coils and thereby cause significant changes in B_0. Because the power consumed by an air-core magnet is proportional to B_0^2, the requirement of a highly stable power supply and effective cooling system makes it very costly to utilize air-core magnets at field strengths above 0.2 Tesla. Resistive magnets operating at higher fields in the 0.3–0.5 Tesla range are designed by using the iron-core configuration composed of resistive wires encompassing two iron pole bars and an iron frame that provides a return path for the magnetic flux.

Superconducting Magnets

In 1911, H. Kamerlingh-Onnes discovered that as temperature decreased to approximetely $4°K$ the resistance of mercury suddenly became zero. This phenomenon of zero resistance at low temperatures is known as *superconductivity*, and materials demonstrating this phenomenon are referred to as *superconductors*. The use of superconductivity was inhibited for several decades by the fact that the utilized materials, known as Type I superconductors, lost their superconductivity as applied magnetic field exceeded a relatively low threshold (typically smaller than 0.1 Tesla). The use of superconducting materials has increased dramatically since the discovery in the late 1950s of new compounds, known as Type II superconductors. The discovered materials can exist in the superconducting state at very high magnetic field strength (on the order of 10 Tesla). Because of this property, Type II superconductors have become the material of choice for high field magnets operating at $B_0 \geq 1$ Tesla. All known superconductors exhibit the superconducting state only at temperatures below a critical temperature, T_c, which is characteristic of the material used. Since conventional Type I and Type II superconductors have very low T_c (normally less than 10–$20°K$), their use usually requires cryogenic systems that employ liquid helium as a coolant. New compounds, discovered in the 1980s and known as High Temperature Superconductors (HTS), have much higher T_c, which makes it possible to use relatively inexpensive liquid nitrogen as a coolant instead of the more expensive liquid helium. However, HTS materials with both high mechanical stability and the ability to maintain the superconducting state at the current densities required for high field magnets have yet to be developed.

Modern superconducting materials for MRI magnets include niobium–titanium (NbTi) and niobium–tin (Nb_3Sn), with NbTi ($T_c = 9.5°K$) being used most often. A schematic of a typical superconducting magnet is shown in Figure 9.2. To maintain the

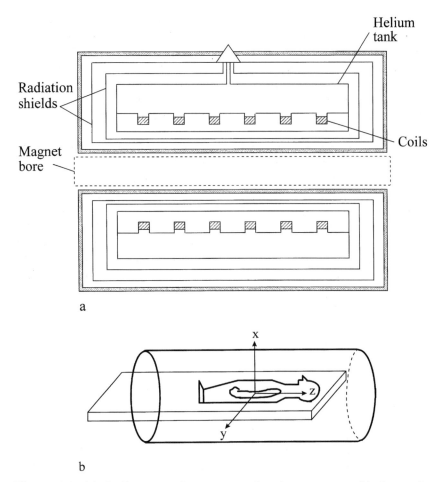

Figure 9.2. (a) A diagram of a superconducting magnet; (b) the main field $\mathbf{B_0}$ in a whole-body superconducting magnet is parallel to the long axis of the patient.

temperature of the NbTi coils below T_c they are placed in a tank with liquid helium. The heat flow to the liquid helium from its surroundings is minimized through the use of two radiation shields mounted outside of the helium tank and kept at low temperatures by an auxiliary cryostat. In some magnets additional thermal isolation of the liquid helium is achieved by employing an outer tank filled with liquid nitrogen.

When activating a superconducting magnet, the current is created initially by an external power supply connected to the superconducting coils through a special switch. While the magnet is being energized, the switch is kept in the normal state by heating. After the operating magnetic field is achieved, the magnet is disconnected from the power supply and the heating of the switch stops. The switch then returns to its superconducting state required to establish a zero resistance path for the operating current in the magnet.

The superconducting state can suddenly disappear as a result of internal heat generated in the magnet coils for a number of reasons. One source of heat may be friction that occurs during small displacements of the superconducting coils caused by Lorentz forces. The transformation to the normal state, known as *quench*, can be quite dramatic because of a rapid (a few seconds) dissipation of the magnetic field energy. The rapid dissipation of energy in a superconducting magnet causes boiling and evaporation of helium and may damage the magnet coils. Nowadays, as a result of the improved design, magnet quenching has become a rare event that occurs without causing permanent damage to the magnet.

Design of Coil Arrangements for Electromagnets

When designing electromagnets (resistive or superconducting), the objective is to devise coil arrangements that provide a specified magnetic field uniformity. To understand the basic approach for minimizing spatial variation in $\mathbf{B_0}$, we consider arrangements of co-axial coils (Figure 9.3). Since in this case the main field $\mathbf{B_0}$ is oriented primarily along the coil axis (shown as the z-axis in Figure 9.3), we can neglect the components of $\mathbf{B_0}$ in the plane perpendicular to the axis. The equation describing the longitudinal component, B_z, in the region free of current is

$$\Delta B_z = 0, \tag{9.1.2}$$

where

$$\Delta = \frac{\partial^2}{\partial x^2} + \frac{\partial^2}{\partial y^2} + \frac{\partial^2}{\partial z^2}.$$

Using spherical coordinates r, ϕ, and θ, we can express a general solution of equation (9.1.2) in the following form:

$$B_z = const + \sum_{n=1}^{\infty} \sum_{m=0}^{n} r^n P_n^m(\cos\ \theta)(A_n^m \cos\ m\phi + B_n^m \sin\ m\phi). \tag{9.1.3}$$

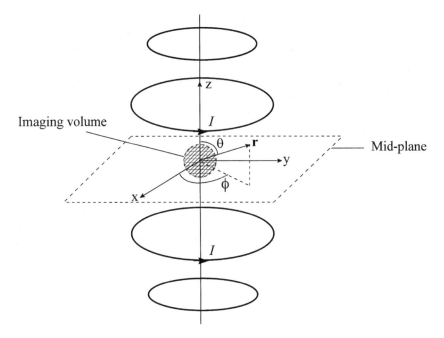

Figure 9.3. A set of coaxial coils forming an electromagnet.

In this equation P_n^m are the associated Legendre polynomials, and A_n^m and B_n^m are arbitrary coefficients. Using the explicit dependence of the lower-order Legendre polynomials on ϕ and θ [4], we obtain that the linear field gradients in the x-, y-, and z-direction are described by the terms with A_1^1, B_1^1, and A_1^0, respectively. The higher-order terms in equation (9.1.3) for which $n > 1$ describe any spatial variations in B_z that are nonlinear as a function of distance from the origin.

To minimize spatial variations in B_z near the origin, we need to devise a configuration of coils for which all coefficients A_n^m and B_n^m with $n = 1, \ldots, N - 1$ are zero, where N indicates a designated order of field uniformity. A simple set of two identical coils placed symmetrically above and below the mid-plane of the magnet and carrying identical parallel currents (Figure 9.3) can be used as a starting point for magnet design. Because of the axial symmetry and the symmetry with respect to the mid-plane of the two-coil magnet, all coefficients A_n^m and B_n^m with $m > 0$ or odd n are zero and the field uniformity in the vicinity of the origin is defined by A_2^0. The field uniformity can be improved by employing the *Helmholtz pair* of coils for which the axial distance between coils equals the coil radius. In this case A_2^0 is zero and the field uniformity is defined by A_4^0. By adding two additional coils it is possible to null A_4^0, leaving the term with A_6^0 as the main contaminant for B_z. The sixth-order contaminant can be eliminated in a system of six coils [1], which is used in practice because it provides sufficiently high field uniformity. In principle, by increasing the number of coils it is possible to null all spatially dependent contaminants of B_z up to any order.

Shimming

Although magnets for MRI are manufactured with great precision to satisfy the design requirements, including coil geometry, symmetric positioning of the coils, etc., small deviations from the specified parameters are practically unavoidable. As a result, the main magnetic field on a MRI scanner typically needs further adjustment, which is done at the imaging site by employing a process known as *shimming*. One approach for shimming, referred to as *passive shimming*, is based on the use of a number of pieces of ferromagnetic material (e.g., iron). Passive shimming normally requires computer analysis to determine the optimal locations and the number of ferromagnetic pieces needed to provide the required magnetic field uniformity. The second approach, known as *active shimming*, utilizes special shim coils (resistive or superconducting) to produce an auxiliary magnetic field $\mathbf{B}_{shim}$, which cancels spatial variations in the main field $\mathbf{B}_0$.

Another objective of shimming is to improve field homogeneity in the imaged specimen, which itself creates a nonuniform magnetic field. In the presence of a specimen, $\mathbf{B}_{shim}$ is usually adjusted interactively based on the acquired signal. Interactive shimming can be performed manually or automatically (i.e., computer-based shimming) by varying the shim currents. After each variation, the signal from a chosen volume is acquired and examined. Initially, the signal rapidly decays due to significant magnetic field nonuniformity in the specimen. As the field uniformity improves, the signal decay becomes slower and the corresponding spectral linewidth decreases. The shimming stops after achieving the minimum linewidth. Alternatively, shimming can be implemented by first calculating the map of the field distribution in the sample and subsequently using a numerical algorithm to determine the optimal shim currents [5–7]. Magnetic field mapping can be performed, for example, by collecting signals during consecutive

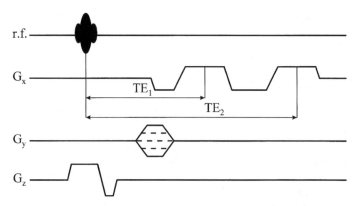

Figure 9.4. Gradient-echo pulse sequence for field mapping.

gradient echoes (Figure 9.4). The phase difference $\delta\phi$ between the image intensities acquired with echo times TE_1 and TE_2, respectively, is given by

$$\delta\phi = \gamma\delta B(TE_1 - TE_2), \qquad (9.1.4)$$

where δB denotes local field inhomogeneity. After δB is determined from the phase difference, the optimal currents for different shim coils can be calculated by minimizing the quantity

$$\int (\delta B + B_{shim})^2 \, dx\, dy\, dz, \qquad (9.1.5)$$

assuming that the magnetic field created by each shim current is known.

Shielding

Magnetic field outside a MR scanner (fringe field) can extend over a significant area and thereby interfere with other devices such as computers, video screens, data storage devices, etc. Consequently, the need to protect the sensitive equipment makes it necessary to minimize the fringe field by utilizing special magnetic screening (shielding) techniques. *Passive shielding* is one common approach used to limit propagation of the main field outside the magnet. It employs specially designed ferrous sheets placed between the magnet and the field-sensitive area. The second approach, referred to as *active shielding*, is based on the use of auxiliary coils. The purpose of the specially designed auxiliary coils is to generate a magnetic field that counteracts the main field produced by the magnet, thus reducing the resultant fringe field.

9.2. GRADIENT COILS

As discussed earlier, spatial localization in MRI is achieved through the use of three orthogonal magnetic field gradients. If the z direction is chosen along the main field B_0, then these gradients can be expressed as $G_x = \partial B_z/\partial x$, $G_y = \partial B_z/\partial y$ and $G_z = \partial B_z/\partial z$. MR imaging generally requires a high degree of gradient linearity throughout the sample volume, rapid switching of the gradients and high gradient strength. An estimate of the minimum gradient strength required for two-dimensional Fourier imaging can be obtained by requiring that the readout time T_s be smaller than T_2^* relaxation time in each voxel. This condition is needed to avoid significant image distortion. The relationship between the readout gradient, G_x, readout time and pixel width in the readout direction, Δx, can be written as

$$\gamma G_x = \frac{2\pi}{T_s \Delta x} > \frac{2\pi}{T_2^* \Delta x}. \tag{9.2.1}$$

At a typical T_2^* of 50 msec and Δx of 1 mm the minimum gradient required is approximately 0.05 gauss/cm. Much higher gradients are required for rapid MR imaging with very short T_s. For example, the readout gradient in EPI performed with T_s of 1 msec and the same spatial resolution as in the example above, is approximately 2.3 gauss/cm. Most commercially available whole-body MR scanners have a maximum gradient strength between 10 and 50 mT/m (1–5 gauss/cm) and a switching time in the 0.1–1 msec range.

In practice, gradient systems are composed of different coil arrangements used to create the x, y, and z gradients. A general mathematical approach used for gradient design employs solutions of the Laplace equation, $\Delta B_z = 0$, with boundary conditions determined by the applied currents, geometry and arrangement of the gradient coils. For the purpose of our discussion, however, we will employ a rather conceptual description of the subject by focusing on basic arrangements of gradient coils.

A commonly used coil configuration for a z-gradient (longitudinal gradient) is the so-called *Maxwell pair*, which consists of two identical coils carrying currents of equal magnitude, I, in opposite directions. These coils are wound on a cylinder coaxial with the magnet coils producing the main magnetic field. The longitudinal component of magnetic field created by a Maxwell pair on the axis of the magnet at distance z from the origin can be expressed as

$$B_z = \frac{\mu_0 a^2 I}{2} \left[\frac{1}{\left[\left(\frac{b}{2} - z \right)^2 + a^2 \right]^{3/2}} - \frac{1}{\left[\left(\frac{b}{2} + z \right)^2 + a^2 \right]^{3/2}} \right],$$ (9.2.2)

where a is the coil radius and b is the distance between the coils. By using the Tailor expansion for B_z one can show that if the distance between the coils equals $a\sqrt{3}$ then the main contaminant of the linear field gradient near the origin is proportional to z^5. Use of additional coils makes it possible to improve the gradient linearity by nulling the fifth-order term. However, adding more coils increases the gradient system inductance. An increase in inductance in turn prolongs the so-called rise time needed to increase the gradient strength from zero to its maximum value. As a result, design of gradient coils involves a compromise between the conflicting requirements of short rise time and a high degree of gradient linearity.

Magnetic field gradients in the x- and y-directions (transverse gradients) are commonly produced by using two identical coil arrangements that are rotated by 90 degrees around the z-axis relative to each other. A frequently used design for transverse gradients, known as *the Golay coils*, is a double-saddle configuration of eight 120° circular arcs (Figure 9.5). The arcs are connected by eight straight segments parallel to the z-axis. Note that a transverse gradient (i.e., $\partial B_z / \partial x$ or $\partial B_z / \partial y$) is created by the currents in the arcs without any contribution from the connecting segments. In practice additional arcs are often added to

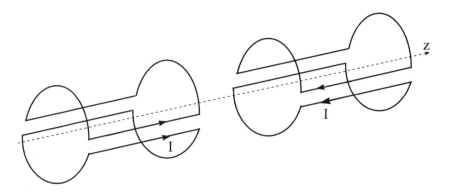

Figure 9.5. Double-saddle gradient coil.

improve the gradient linearity [2]. References [1,2,8] contain detailed considerations of the Golay coils as well as other coil arrangements for the transverse field gradients.

The size of gradient coils is a very important parameter in gradient system design. Consider, for example, the z-gradient produced by a Maxwell pair. By using equation (9.2.2) it is easy to see that near the origin $G_z \propto I/a^2$. This dependence implies that at fixed gradient strength the gradient current I is proportional to a^2. Because the coil resistance is proportional to the coil radius, the voltage across the coil is proportional to a^3 and the power dissipation in the gradient system is proportional to a^5. This indicates that a significant improvement in gradient performance can be achieved by reducing the size of gradient coils.

Eddy Currents

Rapid switching of magnetic field gradients during MR imaging creates large changes in magnetic flux through conducting structures in the magnet (e.g., shim coils, main magnet coils, r.f. coils, etc.). These changes in magnetic flux give rise to eddy currents, which in turn produce both temporally and spatially varying magnetic field. Through their magnetic field, eddy currents cause image distortion and loss of signal. The magnitude, duration, and spatial distribution of eddy currents depend upon various factors including the following: gradient strength and rise time, geometry of the gradient coils, distance between the gradient coils and other conducting structures in the magnet.

One approach for minimizing the effect of eddy currents entails the use of specially designed gradient waveforms that compensate for the magnetic field created by eddy currents [9,10]. To understand the basic idea of the electronic compensation, consider the time dependences of the ideal gradient waveform and the contaminant magnetic field produced by the eddy currents. Since the contaminant field tends to negate the gradient field at the beginning of the pulse, the gradient strength needs to be appropriately increased to provide the necessary overshoot at this stage of the pulse (Figure 9.6). Conversely, the gradient strength at the tail of the pulse needs to be appropriately decreased to compensate for the positive contribution of the eddy currents. In practice the induced eddy currents may require compensation for many milliseconds after the gradient field has changed. Consequently, electronic compensation of eddy currents can prevent the use of compact gradient waveforms making it difficult to achieve short echo and readout times in rapid MR imaging.

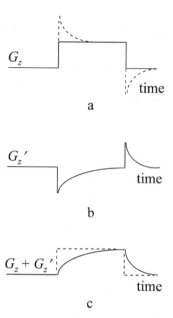

G_z

time

a

G_z'

time

b

$G_z + G_z'$

time

c

Figure 9.6. Schematic illustration of eddy-current compensation: (a) dashed and solid lines indicate the gradient waveforms with and without eddy-current compensation, respectively; (b) contaminant gradient produced by the eddy-currents; (c) dashed and solid lines indicate the resultant gradient with and without eddy-current compensation, respectively.

A better solution to the eddy current problem is to shield the conducting parts of the magnet from the time-varying gradient field. In practice this approach entails the use of the so-called *actively shielded* gradient coils with minimum fringe field [11]. An actively shielded gradient coil actually consists of two coils: an inner (primary) coil, which produces the required gradient field, and an outer (shield) coil, which is designed to cancel the fringe field created by the inner coil. The mathematical framework for the design of actively shielded coils is discussed in [8,11].

9.3. RADIOFREQUENCY COILS

The main functions of r.f. coils are to excite magnetization in a sample and to receive the signal produced by the excited magnetization. These two functions may be performed by using different transmitting and receiving r.f. coils. Alternatively, the same coil may be used both as a transmitter and as a receiver. The requirements for r.f. coils vary depending on the objective of a particular MRI study. For example, when imaging the brain, it may be essential that a utilized coil produce a relatively uniform r.f. field in the imaging volume. On the other hand, if r.f. field uniformity is not critical, it may be desirable to use the so-called surface coils that are built in various sizes and shapes to accommodate the geometry of the sample.

Coil Sensitivity

From the principle of reciprocity [12], it follows that the induced *emf* in the receiver coil in general increases with an increasing ratio of the magnetic field $\mathbf{B_1}$ produced by the coil to the coil current I. When the ratio $\mathbf{B_1}/I$, known as the coil *sensitivity*, is high, a given coil can also be effective as a transmitter. From the Biot and Savart law it follows that

$$\frac{\mathbf{B_1}}{I} = \frac{\mu_0}{4\pi} \int \frac{d\mathbf{l} \times \mathbf{r}}{r^3}. \qquad (9.3.1)$$

In this equation $Id\mathbf{l}$ is a current element, $\mathbf{r}$ is the vector between $d\mathbf{l}$ and the location where $\mathbf{B_1}$ is measured.[2] Equation (9.3.1) shows that the ratio $\mathbf{B_1}/I$ depends only upon the coil geometry. If the z axis is taken in the direction of the main field $\mathbf{B_0}$, we can express the sensitivity $\mathbf{B_1}/I$ as the sum of its transverse component, $\mathbf{B_{1,tr}}/I$, and longitudinal component, $\mathbf{k}B_{1,z}/I$, where $\mathbf{k}$ is the unit vector in the z direction. According to the principle of reciprocity, the induced *emf* in the coil can then be written as

$$emf = - \int_{sample} \frac{\mathbf{B_{1,tr}}}{I} \cdot \frac{\partial \mathbf{M_{tr}}}{\partial t} dV - \int_{sample} \frac{B_{1,z}}{I} \cdot \frac{\partial M_z}{\partial t} dV, \qquad (9.3.2)$$

where $\mathbf{M_{tr}}$ and $\mathbf{k}M_z$ are the transverse and longitudinal magnetizations, respectively.

Following an excitation pulse the transverse magnetization exhibits oscillatory motion with frequency γB_0. Since the inequality $\gamma B_0 T_1 \gg 1$ is normally satisfied in NMR spectroscopy and imaging, the second term on the right side of equation (9.3.2) can be neglected. It is then clear that the coil sensitivity in the transverse plane (i.e., plane perpendicular to $\mathbf{B_0}$) is the quantity that should be maximized when designing r.f. coils, while the longitudinal component of $\mathbf{B_1}$ is

---

[2] Strictly speaking, equation (9.3.1) is valid at low frequencies, when the wavelength of $\mathbf{B_1}$ in air is large compared to the dimensions of the r.f. coil.

inconsequential. In particular, we can conclude that solenoidal coils are not practical for diagnostic MR imaging with superconducting magnets for which $\mathbf{B_0}$ is parallel to the long axis of the patient, because these coils produce mostly a longitudinal r.f. field when encompassing the patient. On the other hand, solenoidal coils are useful when imaging with permanent or resistive magnets that have a main field perpendicular to the long axis of the patient.

Tuning and Matching

The efficiency of a given NMR coil as a transmitter and receiver of electric power depends upon its inductance L, resistance R and capacitance C. In many instances NMR r.f. coils can be described by using equivalent LRC circuits (Figure 9.7a). Suppose that $V_0 e^{j\omega t}$ is the driving *emf* during an excitation pulse. The equation describing the current, I, through the circuit can be written as

$$V_0 e^{j\omega t} = L\frac{dI}{dt} + RI + \frac{q}{C},\qquad (9.3.3)$$

where q is the charge stored by the capacitor. It is easy to verify that the induced current is given by

$$I = \frac{V_0 e^{j(\omega t - \phi)}}{\sqrt{R^2 + \left(\omega L - \dfrac{1}{\omega C}\right)^2}},\qquad (9.3.4)$$

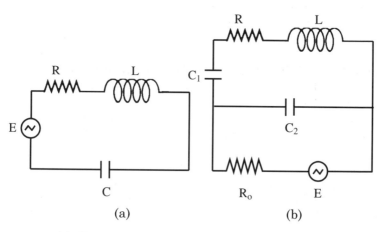

Figure 9.7. (a) Circuit diagram for a typical r.f. coil; (b) impedance matching scheme.

where

$$\tan\phi = \frac{\omega L - \dfrac{1}{\omega C}}{R}.$$

To effectively generate the required B_1 field, it is generally desirable to have the maximum current amplitude at the frequency $\omega_0 = \gamma B_0$. In practice this can be achieved by adjusting (tuning) the coil capacitance in such a way that

$$\frac{1}{LC} = \omega_0^2. \qquad (9.3.5)$$

This condition implies that the r.f. coil resonates at the Larmor frequency of the nuclei. The width of the resonance at half-height, $\delta\omega$, can be expressed as $\delta\omega \approx \sqrt{3}\omega_0/Q$, where Q, known as the coil *quality factor*, is given by

$$Q = \frac{\omega_0 L}{R}. \qquad (9.3.6)$$

As discussed previously (see Chapter 5) the signal-to-noise ratio is inversely proportional to the square root of R, and since a high Q indicates a low resistance R, it is desirable to design r.f. coils with high Q values. Note that because the effective coil resistance R is given by the sum of the sample and the coil resistances [13], the quality factor for a "loaded" coil (i.e., sample is present) will be significantly smaller than the quality factor measured without a sample if the coil resistance is quite small compared to the sample resistance.

When operating, the NMR r.f. coil is connected with the transmitter and receiver electronics via a coaxial cable (transmission line). To transmit electric power to and from the coil most efficiently, the impedance of the r.f. coil must be equal to the impedance of the transmission line. The line impedance is typically $50\,\Omega$ resistive. Because the r.f. coil resistance is usually less than $50\,\Omega$, it is necessary to use impedance matching networks to ensure that the r.f. coil is equivalent to a $50\,\Omega$ resistance when operating at the resonance frequency ω_0. The transformation of the coil impedance to $50\,\Omega$ can be achieved through the use of various matching schemes. One such scheme employing external capacitors C_1 and C_2 is shown in Figure 9.7b. It can be shown that for this scheme the required matching of the coil impedance occurs when

$$\omega_0 L - \frac{1}{\omega_0 C_1} = [R(R_0 - R)]^{1/2} \quad \text{and} \quad \frac{1}{\omega_0 C_2} = R_0(R_0/R - 1)^{-1/2}. \quad (9.3.7)$$

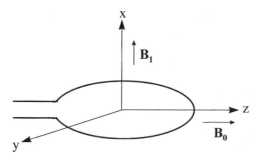

Figure 9.8. Circular surface coil.

Coil Types

Among the various types of r.f. coils used for imaging, surface coils are the most basic. Surface coils offer the advantage of high sensitivity when used for imaging superficial structures in a specimen. An example of a surface coil is a circular loop of conducting wire (Figure 9.8). The B_1 field along the coil axis (taken as the x axis) is given by

$$B_1 = \frac{\mu_0 I a^2}{2(a^2 + x^2)^{3/2}}, \tag{9.3.8}$$

where a is the coil radius. By examining this equation we can make several important conclusions. First, an r.f. field produced by a single surface coil is very inhomogeneous, making it difficult to obtain a good slice profile during MR imaging. As a result, surface coils are used primarily as receiving coils. Second, the sensitivity of a given surface coil rapidly decreases with increasing distance from the center of the coil. This explains why surface coils are used for imaging only when they can be placed in close proximity to the regions to be imaged. Third, according to equation (9.3.8) the sensitivity at the center of a circular surface coil is inversely proportional to the coil radius. Therefore, small surface coils placed on the surface of a sample provide a higher sensitivity near the surface compared to that of large coils. On the other hand, the sensitivity of small surface coils decreases more rapidly with increasing depth from the surface than the sensitivity of large coils (Figure 9.9).

When used for imaging, a single surface coil normally provides acceptable B_1 homogeneity only in the regions that have dimensions smaller than or comparable to the diameter of the coil. Consequently,

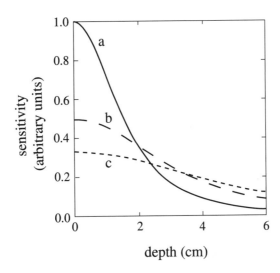

Figure 9.9. Sensitivity vs depth for circular surface coils with different radii: (a) $a = 2$ cm; (b) $a = 4$ cm; (c) $a = 6$ cm.

there exist two options for imaging a large region with surface coils: (a) use of a single large coil; (b) use of an array of relatively small coils, known as the *phased array coil* [14]. A phased array coil consists of a number of noninteracting surface coils (e.g., 4 or 8) that simultaneously receive NMR signals. Although each surface coil effectively images only a small region within the field-of-view, individual images produced by an array of coils can be combined together to obtain an image of a large area with uniform sensitivity and with SNR higher than that of a single large surface coil.

For many applications of MRI such as body or head imaging, a homogeneous transverse B_1 field is essential. A transverse B_1 can be produced by a current flowing on the surface of an infinitely long cylinder parallel to its axis. In this case, the condition required for a uniform B_1 is that the current density is proportional to $\sin \phi$ (or $\cos \phi$), where ϕ is the azimuthal angle [15]. The sinusoidal distribution of current density can be approximated by using a number of discreet wires in the design known as the *saddle coil*. A typical saddle coil (Figure 9.10(a)) consists of four linear segments and four 120° circular arcs on a cylindrical surface. This design approximates the $\sin \phi$ distribution by having two linear segments at 60° and 120° carrying equal currents in one direction and two other segments at 240° and 300° carrying equal currents in the opposite direction. The sinusoidal

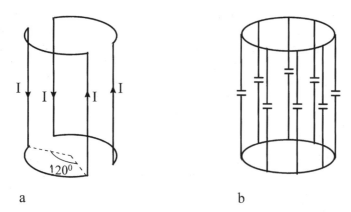

Figure 9.10. (a) Saddle coil; (b) birdcage coil.

current density can be more accurately approximated by using the so-called *birdcage coil*, which includes a large number of closely spaced linear segments (e.g., 16) uniformly distributed on a cylindrical surface [16]. B_1 homogeneity in the birdcage coil (Figure 9.10(b)) is much higher than in a saddle coil of comparable size. It is interesting that the r.f. field in a birdcage coil can be made circularly polarized by simultaneously exciting two orthogonal modes corresponding to $\sin \phi$ and $\cos \phi$ distributions of current. When used for signal reception, these modes can be combined together after one of them is phase shifted by 90 degrees. This doubles the NMR signal while the noise increases only by a factor of $\sqrt{2}$ because the noise signals associated with the two modes are not correlated. Consequently, the SNR increases by a factor of $\sqrt{2}$ [17]. Application of circularly polarized field for signal excitation is discussed in the following section.

Quadrature Excitation

As discussed previously (see Chapter 1), excitation of nuclear magnetization in a static magnetic field $\mathbf{B_0} = \mathbf{k}B_0$ can be performed by a linearly polarized r.f. field $\mathbf{B_1} = \mathbf{i}B_1 \cos \omega t$, where $\mathbf{i}$ and $\mathbf{k}$ are unit vectors in the x- and z-directions, respectively. A linearly polarized field can be viewed as the sum of two components rotating around the z-axis at the same frequency but in the opposite directions. Since the transverse magnetization in a sample is predominantly created by the component of $\mathbf{B_1}$, which rotates in the direction of Larmor precession of spins, only half of the r.f. power is used effectively for excitation.

Alternatively, excitation of nuclear magnetization can be achieved by using a circularly polarized r.f. field rotating in the direction of Larmor procession of spins. Since in this case, known as the *quadrature excitation*, all r.f. power is effectively utilized, the use of a circularly polarized magnetic field lowers the required r.f. power as well as the power deposited in tissue by a factor of 2.

REFERENCES

[1] K. G. Dobson. "Magnet design and technology for NMR imaging," *NMR in Medicine: The Instrumentation and Clinical Applications*, S.R. Thomas and R.L. Dixon, eds., American Institute of Physics, NY, 85 (1986).

[2] S.R. Thomas. "Magnets and gradient coils: types and characteristics," *The physics of MRI*, M.J. Bronskill and P. Sprawls, eds., American Institute of Physics, NY, 56 (1993).

[3] W.H. Oldendorf. "A comparison of resistive, superconductive and permanent magnets," *Magnetic Resonance Imaging, 2nd Edition*. Volume II: Physical Principles and Instrumentation, C.L. Partain, R.R. Price, J.A. Patton et al. eds., W.B. Saunders Company, Philadelphia, PA, 1133 (1988).

[4] M. Abramowitz, I. Stegun. *Handbook of Mathematical Functions*. Dover Publications, New York (1972).

[5] M.G. Prammer, J.C. Haselgrove, M. Shinnar, J.S. Leigh. "A new approach to automated shimming," *J. Magn. Reson.*, **77**, 40 (1988).

[6] J. Tropp, K.A. Derby, C. Hawryszko. "Automated shimming of B_0 for spectroscopic imaging," *J. Magn. Reson.*, **85**, 244 (1989).

[7] P. Webb, A. Macovski. "Rapid, fully automated, arbitrary-volume *in vivo* shimming," *Magn. Reson. Med.*, **20**, 113 (1991).

[8] R. Turner. "Gradient coil design: A review of methods," *Magn. Reson. Imag.*, **11**, 903 (1993).

[9] D.J. Jensen, W.W. Brey, J.L. Delayre, P.A. Narayana. "Reduction of pulsed gradient settling time in the superconducting magnet of a magnetic resonance instrument," *Med. Phys.*, **14**, 859 (1987).

[10] J.J. Van Vaals, A.H. Bergman. "Optimization of eddy-current compensation," *J. Magn. Reson.*, **90**, 52 (1990).

[11] P. Mansfield, B. Chapman. "Multishield active magnetic screening of coil structures in NMR," *J. Magn. Reson.*, **72**, 211 (1987).

[12] D.I. Hoult, R.E. Richards. "The signal-to-noise ratio of the nuclear magnetic resonance experiment," *J. Magn. Reson.*, **24**, 71 (1976).

[13] D.I. Hoult, P.C. Lautebur. "The sensitivity of the zeugmato-graphic experiment involving human samples," *J. Magn. Reson.*, **34**, 425 (1979).

[14] P.B. Roemer, W.A. Edelstein, C.E. Hayes, S.P. Souza, O.M. Mueller. "The NMR phased array," *Magn. Reson. Med.*, **16**, 192 (1990).

[15] D.E. Lobb. "Properties of some useful two dimensional magnetic fields," *Nucl. Instrum. Methods*, **64**, 251 (1968).

[16] C.E. Hayes, W.A. Edelstein, J.F. Schenck, O.M. Mueller, M. Eash. "An efficient, highly homogeneous radiofrequency coil for whole-body NMR imaging at 1.5 Tesla," *J. Magn. Reson.*, **63**, 622 (1985).

[17] C.N. Chen, D.I. Hoult, V.J. Sank. "Quadrature detection coils: a further $\sqrt{2}$ improvement in sensitivity," *J. Magn. Reson.*, **54**, 324 (1983).

Appendix

A. PHASE-SENSITIVE DETECTION

As mentioned earlier, the observed NMR signal typically consists of different components in the narrow range of frequencies $\omega_0 \pm \delta\omega$, where $\delta\omega \ll \omega_0$. The direct sampling of NMR signal is difficult because it requires a very high sampling rate during analog-to-digital conversion. To lower the required sampling rate the signal can be shifted down in frequency by ω_0 prior to analog-to-digital conversion while the phases and relative amplitudes of different signal components within the bandwidth $2\delta\omega$ are preserved. In practice such reduction in the signal frequency is commonly achieved by using two identical circuits known as *phase-sensitive detectors* (shown as PSD_1 and PSD_2 in Figure A).

To explain the operation of these detectors, we consider a simple case when the observed signal, S, is given by

$$S = S_0 \cos(\omega t + \phi), \qquad (A.1)$$

where S_0, ω and ϕ are the amplitude, frequency and phase of the signal, respectively. Initially the signal is amplified by a low-noise amplifier. The amplified signal is then divided equally between the two phase-sensitive detectors. In PSD_1 the incoming signal is multiplied by $\cos\omega_0 t$. The output signal, S_1, can therefore be expressed as

$$S_1 \propto S_0 \cos[(\omega - \omega_0)t + \phi] + S_0 \cos[(\omega + \omega_0)t + \phi]. \qquad (A.2)$$

Because the frequency of the NMR signal is close to ω_0 (i.e., $|\omega - \omega_0| \ll \omega_0$), the term $S_0 \cos[(\omega + \omega_0)t + \phi]$ describes the rapidly

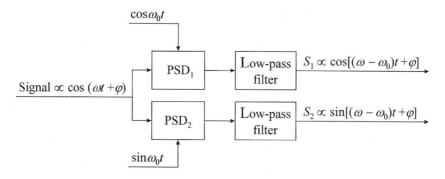

Figure A. A diagram of the phase-sensitive signal detection.

varying signal component at the frequency of $\omega + \omega_0 \approx 2\omega_0$. In contrast, the term $S_0 \cos[(\omega - \omega_0)t + \phi]$ describes the slowly varying component of the output signal. The rapidly varying component is eliminated by low-pass filtering of S_1.

In PSD_2 the incoming signal is multiplied by $\sin \omega_0 t$. The output signal, S_2, is therefore given by

$$S_2 \propto -S_0 \sin[(\omega - \omega_0)t + \phi] + S_0 \sin[(\omega + \omega_0)t + \phi]. \tag{A.3}$$

Again the rapidly varying signal, proportional to $\sin[(\omega + \omega_0)t + \phi]$, is removed by low-pass filtering of S_2. The resulting output signals S_1 and S_2 can be combined into one complex signal

$$S_1 - jS_2 \propto S_0 e^{j(\omega - \omega_0)t + j\phi}. \tag{A.4}$$

Let us now consider phase-sensitive detection of the NMR signal in two-dimensional MR imaging. Neglecting the effect of magnetic field inhomogeneities, the NMR signal from a sample can be written as

$$S \propto \iint M_0 e^{-TE/T_2} \cos(\omega_0 t + k_x x + k_y y + \phi) \, dx \, dy, \tag{A.5}$$

where M_0 is the equilibrium nuclear magnetization, TE is the echo time, k_x and k_y are dependent upon the readout and phase-encoding gradients, respectively. Taking into account equations (A.2)–(A.4) it is easy to show that after phase-sensitive detection and low-pass filtering, the complex output signal $S_1 - jS_2$ is given by

$$S_1 - jS_2 = \xi \iint M_{xy} e^{jk_x x + jk_y y} \, dx \, dy, \tag{A.6}$$

where $M_{xy} = M_0 e^{-TE/T_2 + j\phi}$ and ξ is an arbitrary constant.

B. IMAGE DISPLAY

Images are typically displayed on the screen of an image device by using a number of screen elements (e.g., 512 × 512), referred to as display pixels. The screen location of a pixel representing a given element from a two-dimensional data array is defined by the address of the element within the array. The pixel color is defined by the value of the element. There are several standard systems for reproducing color. One such system, known as the RGB (red, green and blue), is often implemented using $2^8 = 256$ shades of each main color (i.e., red, green and blue). Consequently, the total color palette including shades of gray consists of 2^{24} combinations of the main colors.

Diagnostic MR images of various anatomical structures are normally displayed by using a number (e.g., 256) of different shades of gray chosen according to the utilized relationship between intensity and brightness. Among countless possibilities for mapping intensity values to the displayed brightnesses, a linear relationship between intensity and brightness is used most often. In various instances image contrast can be enhanced if a limited range of intensity values is converted to the full range of brightnesses. In this approach intensity values smaller than an arbitrary chosen lower limit are typically represented by the black color, intensity values greater than an arbitrary chosen upper limit are represented by the white color and the intermediate intensity values are displayed by using different shades of gray. Usually the upper and lower limits are adjusted manually or automatically to enhance contrast between various image structures.

Index